Molecules, Medicines and Mischief

A Year on the Chemical Trail around Cambridge University Botanic Garden

Gwenda Kyd and Mo Sibbons

Vervain Publishing

Published by
Vervain Publishing
39 Regatta Court, Oyster Row
Cambridge, CB5 8NS
UK

A catalogue record for this book is available from the British Library.
ISBN 978-0-9928998-0-6

Designed and typeset by Dora A. Kemp

Printed in Great Britain by Short Run Press, Exeter

Contents

Preface

This publication would not have been possible without the efforts of many people and we would like to extend our thanks to everyone who contributed. The work is broadly based on the Chemicals from Plants Trail at Cambridge University Botanic Garden (CUBG). The Trail was a collaboration between CUBG and Cambridge Crystallographic Data Centre (CCDC). From CUBG, Juliet Day, Dr. Tim Upson and Professor John Parker contributed to the devising and realisation of the Trail along with Dr. Frank Allen, Dr. Gary Battle and one of us (GK) from CCDC. When the Trail launched in May 2012 it contained 22 plants.[1,2,3,4] In the course of writing this book the Trail was extended to 26 plants and we'd like to thank Juliet Day for updating the online version.[5]

Some photos included in the book were kindly contributed by Juliet Day (CUBG), Jenny Humphrey (Citizens for a Liveable Cranbrook), Lynher Dairies (who manufacture Cornish Yarg cheese, http://www.lynherdairies.co.uk/) and the many contributors to Pixabay. Dr. Norman Bennett of Allicin International Limited is thanked for providing access to papers on stabilised allicin.

We'd like to thank Dora A. Kemp for designing and typesetting this book. The project has been interesting and rewarding but costly and we'd like to thank Rick Davies, Barry Peachey and other contributors who wish to remain anonymous for their contributions to meeting the costs involved. Finally, the patience and input of Mary Kyd and Dr. Seth Wiggin who edited text and Eirlys Neale who proof-read it, are gratefully acknowledged.

References

1. G.M. Battle, G.O. Kyd, C.R. Groom, F.H. Allen, J. Day, T. Upson, (2012); *Journal of Chemical Education*, 89, 1390–4, doi: 10.1021/ed300065s
2. G. Kyd, (2012); *Crystallography News*, 123, 16–18, http://crystallography.org.uk/wp-content/uploads/2011/06/BCA-News-December-2012.pdf
3. G. Kyd, (2012); *ACA Reflexions*, 4, 12, http://www.amercrystalassn.org/documents/newsletterarchive/winter.WEB.pdf
4. G. Kyd, (2012); *Herbs*, 37(3), 19–21
5. The online Trail including a map showing the location of the 26 plants in CUBG is available at www.bit.ly/CCDCTrail

Further Information

Disclaimer: The text is intended for entertainment purposes. Plants should not be used for medical purposes without seeking expert advice. The medicinal activities of plant material or chemical compounds are described for information only and this should not be taken to imply that they are or will be available for human use now or in the future.

For more information on herbal medicine it is recommended that one of the professional associations are consulted (e.g. European Herbal and Traditional Medicine Practitioners Association, http://ehtpa.eu/index.html or the National Institute of Medical Herbalists, http://www.nimh.org.uk/). More information on homoeopathy and a list of registered homoeopaths can be found at http://www.britishhomeopathic.org/, more information on Bach Flower Remedies can be obtained at www.cambridge-bach.co.uk and a list of registered practitioners and further information from the Bach Centre at www.bach-centre.com.

Notes on chemical diagrams, structures and references: For each chemical compound named in the section headings, a chemical diagram is included. This is a two dimensional representation of the chemical structure. By convention, all unlabelled atoms are carbon (C) and these are assumed to have sufficient hydrogen atoms (H) attached to give standard valency (i.e. 4 single connections to other atoms). Non-carbon atoms such as oxygen (O),

nitrogen (N) and sulfur (S) are shown with any attached H atoms.

Many techniques can be used to study extracts from plants or the individual compounds isolated. One of these is X-ray crystallography which models the positions of individual atoms in a crystal and allows connections and interactions between them to be studied. When atoms are connected together they form a molecule. Looking at a collection of structures can allow the active part of a molecule to be identified and by extension, the activity maximised. All of the featured compounds in this book shown in the chemical diagrams have been studied by X-ray crystallography in the solid state.

References are included in each section. These are a mixture of websites, news articles and scientific journal articles. Lists are not comprehensive but are included to provide a starting point for those wishing further information.

Photo credits: Photos are by Mo Sibbons with the exception of the following:
Pages 1, 52, 62(right), 68(left), 81, 92, 103(left): Juliet Day
Page 2: Public Domain courtesy of Los Angeles County Museum of Art, www.lacma.org
Pages 3, 26, 31, 32 (right), 33, 37, 38, 44(right), 50, 56, 63, 68(right), 76, 91(left), 91(right), 93, 97, 98, 111, 112, 116, 118, 121, 127, 128, 129, 133, 134, 135, 147, 148: Pixabay.com and contributors
Page 15: ©Science Museum/Science & Society Picture Library
Page 19: Stephen Ausmus/ARS-USDA
Page 28: ©Antonio Gravante/Shutterstock.com
Page 34: ©Rob Stark/Shutterstock.com
Page 39: ©Indigoiris/Shutterstock.com
Page 44 (left): ©Roberto Tetsuo Okamura/Shutterstock.com
Page 45: Lynher Dairies, www.lynher.co.uk
Page 51: ©Andrei Rybachuk/Shutterstock.com
Page 62(left): ©Nitr/Shutterstock.com
Page 70: ©foto76/Shutterstock.com
Page 74: Jenny Humphrey
Page 79: ©PunyaFamily/Shutterstock.com
Page 87: ©Steffen Hauser
Page 94: ©Stacey Lynn Payne/Shutterstock.com
Page 103(right): ARS-USDA
Page 104: ©Africa Studio/Shutterstock.com
Page 105: Scott Bauer/ARS-USDA
Page 106: ©vblinov/Shutterstock.com
Page 117: ©Bildagentur Zoonar GmbH/Shutterstock.com
Page 122(left): ©ConstanzeK/Shutterstock.com
Page 122(right): ©J. Henning Buchholz/Shutterstock.com
Page 154: ©Nikitin Victor/Shutterstock.com
Rear flap (upper): Tracey van den Ban

Introduction

For as long as mankind has existed, we have looked to the plants growing around us as a source of food, medicines and materials for almost every aspect of human life. Some of these uses have been superseded but many populations have limited access to these technological advances. Even in more wealthy countries, there has been an increase in the popularity of traditional plant-based medicine and a growing interest in the use of natural products as a green alternative to synthetic chemicals. In the 21st century, plants are every bit as important to our physical and mental well-being as they have ever been.

Plants contain a large number of different chemical compounds. Many of these are present to deter predators or attract pollinators. Others act as anti-oxidants, are involved in temperature control or govern the plant's response to external factors.

The study of these compounds can give us a clearer understanding of how the plant kingdom operates but, also, how plants and their chemical components can be harnessed to maximise the benefit to us. One example of the latter is in medicine. Estimates suggest that around 11% of the 252 medicines considered basic and essential by the World Health Organisation (WHO) are obtained directly from flowering plants.[1] Taking into account the ideas obtained from the study of novel structures from plants, up to

50% of drugs approved in the 30 years prior to 2012 come directly or indirectly from the natural world.[1] Notable success stories include paclitaxel, an anti-cancer drug identified in yew trees and artemisinin, an anti-malarial compound from sweet wormwood. Around 80% of drugs of plant origin are used to treat conditions which correspond to the traditional use of the plant illustrating the current and continuing relevance of traditional knowledge.[2] A recent example of this is ingenol mebutate which can be used to treat some forms of skin cancer. This is obtained from the plant *Euphorbia peplus*, commonly known as cancer weed, which is used in Australia as a traditional remedy for skin cancer. Improved utilisation of plants can also contribute to solving issues such as food shortages, global warming and water and air pollution.

The 26 plants which form the Chemicals from Plants Trail have a multitude of uses and provide examples of the good, the bad, and the sometimes surprising applications of plants and plant-derived compounds.

References

1. C. Veeresham, (2012); *Journal of Advanced Pharmaceutical Technology and Research*, 3(4), 200–1, doi: 10.4103/2231-4040.104709
2. D.S. Fabricant, N.R. Farnsworth, (2001); *Environmental Health Perspectives*, 109(Suppl.1), 69–75, PMCID: PMC1240543

1. Walnut (*Juglans* spp.) and Juglone

The most common members of the genus *Juglans* are English (or Persian) walnut (*Juglans regia*) and black walnut (*Juglans nigra*). *Juglans regia* originated in Persia (Iran) but was commonly grown in England and became known as English walnut. The name walnut comes from the Old English *walh-hnutu*, foreign nut, while the genus name means Jupiter's acorns or nuts. Juglone is one of a number of phenolic compounds produced by walnut trees.

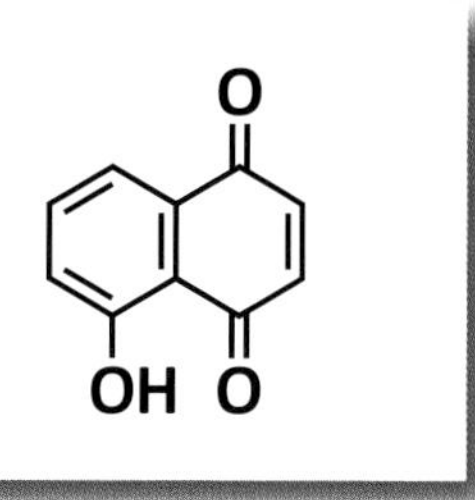

Walnut wood from the slow-growing trees was popularly used to make furniture in the late 17th century, with wood imported from France.[1]

Queen Anne style furniture, popular before, during and after the reign of Queen Anne (1702–1714), was commonly made of walnut. However, a severe winter in France in 1709 damaged or killed many trees and export was banned by the French in 1720. British furniture makers had to find new woods to use and mahogany became popular. Wood from black walnut trees was also used to make aircraft propellers during World War I.

Wood from walnut tree burrs (where the tree grows in a deformed manner) is highly prized for wood-turning and inlaying. Veneers are used in the interior of prestige cars e.g. on dashboards. Walnut flowers, stems, young leaves and stalks are used to make a Bach Flower Remedy to protect from change and unwanted outside influences.

The shells (or husks) of walnuts are used for cleaning bricks, wood, stone and metal. The use of walnut shell grit to clean aircraft engines was discontinued after a crash involving a US Army Chinook helicopter at Mannheim airshow in 1982. The crash which killed 46 people was attributed to an accumulation of the grit. Shells do not absorb water and resist the growth of bacteria and mould making them suitable for use as pet bedding. Walnut shell can be ground to different grain sizes from fine flour to coarse. As a result, it can be used as a filler in many products including paints, adhesives, ceramics, dynamite and livestock feed, adding volume and reducing cost. It is also used in cosmetic products for exfoliation and as a scrub. In some face powders, walnut shell flour is also used to absorb oil.[2]

Walnuts have a cultural significance in some parts of China. Pairs of walnuts are rotated in the palm of the hand both to stimulate circulation and as a status symbol. Symmetrical, large, old walnuts are highly prized and collectable with some selling for thousands of pounds. Street sellers engage buyers in walnut gambling or *du he tao*. The value of the pair of walnuts bought isn't known until the outer shells are opened. Pairs can be valuable or worthless depending on the size, type, grain, shape and similarity between them.[3]

Walnut kernels of all species can be eaten and China, Iran and California are the largest producers. Proper storage is essential to avoid the growth of fungal moulds which produce highly toxic and carcinogenic aflotoxins. Unlike other nuts, walnuts contain mainly polyunsaturated fatty acids such as α-linoleic acid and linoleic acid. Their unique fatty acid content appears beneficial to human health, causing a lowering in cholesterol levels without significant weight gain.[4] α-Linoleic acid is an example of an omega-3 fatty acid. These have been linked to the reduction of inflammation and joint pain, reduction of size and inhibition of growth of prostate tumours and prevention of breast cancer and heart disease.[5]

Walnuts are rich in anti-oxidants such as myricetin to protect the kernels from oxidation which would cause them to turn rancid. They are higher in anti-oxidants than other nuts and, as they are eaten raw, don't suffer the loss of anti-oxidants other nuts do when roasted. Kernels are nutrient rich containing high-quality protein, fats, fibre, calcium, iron, phosphorous, zinc, vitamin B6 and melatonin. Eating as few as 7 walnuts a day can give walnut-associated health benefits.[6]

Walnut oil can be used in salad dressing, for frying and as lamp oil. It can also be used as a drying oil, mixed with pigments to produce oil paints. A drying oil hardens in contact with air and walnut oil does not cause a yellow tinge like linseed oil. Walnut oil was commonly used by Renaissance painters such as Leonardo da Vinci, but the oil has to be carefully stored to prevent it turning rancid.

The Doctrine of Signatures was an important idea in the development of traditional medicine from the Middle Ages. Paracelsus (1493–1541),

a Swiss physician, was an influential advocate.[7] The Doctrine suggested that when the Earth was designed, plants were marked to suggest what they could be used for e.g. the leaves of lungwort, *Pulmonaria officianalis*, look like the inside of lungs and were, therefore, used to treat lung diseases. As walnuts resemble brains, they were traditionally used to treat conditions of the brain and head and by extension, the shell used to treat wounds to the head and hair-loss.

There are numerous other uses of walnut trees in traditional medicine. The bark and leaves have laxative, alterative (i.e. change processes within the body), astringent and detergent properties and were used to treat skin diseases such as scrofula ('the king's evil'), herpes, eczema and ulcers. Walnut was sometimes referred to as 'vegetable arsenic' as it was used to treat conditions such as eczema and other skin diseases traditionally treated with arsenic.[5] The husk and peel were used as a sudorific (to cause sweating). Macerating the husks and leaves in warm water produced a bitter liquid which was used to destroy worms on grass without damage to the grass itself.[7] Juice of green husks boiled with honey was used as a gargle for sore mouths or throats or stomach inflammation and the nuts bruised with honey were used for ear inflammation. A piece of green husk could be put into a hollow tooth to ease inflammation. The oil was used to expel wind and treat colic.[5]

Unripe walnuts are sometimes eaten pickled with meats or cheese and the pickling vinegar can be used as a gargle to treat sore throats.[8] Raw walnuts are used in both savoury and sweet dishes. Green walnuts are used to make the Italian liqueur Nocino while walnuts and hazelnuts are used to make the liqueur Nocello. Walnut husks produce a brown dye which can be used to dye hair or fabric. It does not require a fixative (mordant) and will stain the hands if gloves aren't worn. The husks contain a number of phenolic compounds which can irritate the skin as well as staining. These include juglone, coumaric acid, regiolone and vanillic acid. Juglone itself is used to dye fabrics, as an ink and as an orange-brown colouring agent in the food and cosmetic industries. Juglone, betulinic acid and regiolone have also been isolated from the stem and bark of *Juglans regia*.[9]

Juglone (5-hydroxy-1,4-naphthoquinone) is an allelopathic compound. It is leached into the soil deterring other plants from growing nearby so the available resources are preserved for the walnut tree. However, some plants and trees including birch and maple are immune. Juglone also protects the tree against most herbivores. Some insects can detoxify juglone, however, and convert it to non-toxic 1,4,5-trihydroxy-naphthalene in the gut.[10] Juglone also shows anti-microbial[11] and anthelmintic properties (i.e. it kills parasitic worms).

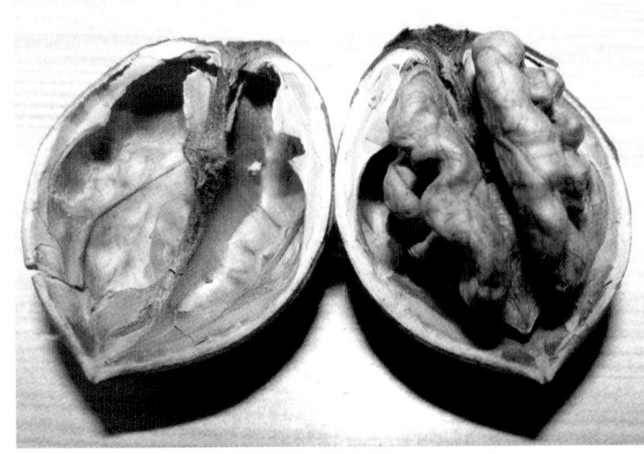

Juglone can be used as a pH indicator, i.e. it is a different colour depending on the acidity of a solution.[12] It is one of the compounds responsible for a depressant action of walnut husks on the activity of some marine life, rats and mice. This action has led to the illegal use of husks in parts of USA to immobilise fish. They lose equilibrium and awareness and can eventually be caught by hand. Juglone itself produces a different quality of sedation, suggesting that in walnut husks it acts synergistically with other compounds.[13] Juglone has shown activity against several types of cancer including gastric cancer,[14] melanoma and leukaemia and it also shows a preventative effect.

References

1. P. Brown, (2011); *The Guardian*, http://www.theguardian.com/environment/2011/apr/25/specieswatch-english-persian-walnut [Accessed 15.12.13]
2. A. Georgalas, T.C. Harris, (1981); *US Patent*, US 4279890
3. Y. Lu, (2012); *Global Times*, http://www.globaltimes.cn/content/738672.shtml [Accessed 16.12.13]
4. D.K. Banel, F.B. Hu, (2009); *American Journal of Clinical Nutrition*, 90(1), 56–63, doi: 10.3945/ajcn.2009.27457
5. T. Breverton, (2011); *Breverton's Complete Herbal*, Quercus Publishing Ltd, ISBN 978-0-85738-336-5, pp 352–3
6. American Chemical Society Press Release, (2011); http://www.acs.org/content/acs/en/pressroom/newsreleases/2011/march/walnuts-are-top-nut-for-heart-healthy-antioxidants.html [Accessed 16.12.13]
7. http://www.sciencemuseum.org.uk/broughttolife/techniques/doctrine.aspx [Accessed 16.12.13]
8. Mrs M. Grieve, (1973 edition); *A Modern Herbal*, Merchant Book Company Ltd., ISBN 1904779018, pp 842–5
9. S.K. Talapatra, B. Karmacharya, S.C. De, B. Talapatra, (1988); *Phytochemistry*, 27(12), 3929–32, doi: 10.1016/0031-9422(88)83047-4
10. R. Piskorski, S. Ineichen, S. Dorn, (2011); *Journal of Chemical Ecology*, 37(10), 1110–6, doi: 10.1007/s10886-011-0015-4
11. A.M. Clark, T.M. Jurgens, C.D. Hufford, (1990); *Phytotherapy Research*, 4(1), 11–14, doi: 10.1002/ptr.2650040104
12. S.S. Sawhney, N.C. Trehan, (1980); *Indian Journal of Chemistry*, 14A, 295
13. B.A. Westfall, R.L. Russell, T.K. Auyong, (1961); *Science*, 134(3490), 1617, doi: 10.1126/science.134.3490.1617
14. Y.-B. Ji, Z.-Y. Qu, X. Zou, (2011); *Experimental and Toxicologic Pathology*, 63(1–2), 69–78, doi: 10.1016/j.etp.2009.09.010

January

1

2

3

4

5

6

7

January

8

9

10

11

12

13

14

2. Willow (*Salix* spp.) and Aspirin

Willow was used medicinally by the ancient Egyptians as shown in the Ebers papyrus, dating from around 1500 BC.[1] In Europe, extracts from the bark have been used to treat pain and inflammation since at least the time of Hippocrates (400 BC). Willows are fast-growing trees commonly found growing near water. The genus name *Salix* comes from the Celtic *sallis*, meaning near water.[2] In the 1750s, Reverend Edward Stone (1702–1768) was suffering from agues (a collection of symptoms including fever, probably due to malaria). Walking near Chipping Norton, he decided to try chewing a small piece of willow bark. Its bitterness was reminiscent of that of cinchona bark, the source of quinine, a well-known treatment for malaria, so Stone investigated further. He tested the powdered bark on 50 patients and reported powerful astringent properties and beneficial effects on agues in 1763, in a letter to the president of the Royal Society, Lord Macclesfield.[3] His use of willow was influenced by the Doctrine of Signatures

as willow trees grow in areas where agues (and rheumatism) were common.

Willow bark contains the compound salicin which is readily oxidised to salicylic acid. Salicylates, ions from salicylic acid, are found throughout the willow tree and also in the wildflower meadowsweet (*Filipendula ulmaria*). Different species of willow contain different amounts with white willow (*Salix alba*) abundant and weeping willow (*Salix babylonica*) containing little. Salicylates are leached into the soil around the willow tree and prevent growth of other plants which would compete with the tree for resources – another example of allelopathy. The astringent properties of willow bark are due to the presence of tannins.

White willow wood is used to make cricket bats (from cricket-bat willow, *Salix alba* var. *caerulea*) and can be used in artificial limbs, for basket-making and fencing. The supply of willow for the cricket bat industry is concentrated in East Anglia. As the wood was also used in the gallows, it was sometimes considered unlucky to cut the trees down. Fen residents were, therefore, reluctant to take trimmings from trees felled for cricket bat making, even when offered for free. Trees are felled at 10 years old and each makes about 32 bats.

Willow twigs, called osiers, are used to make baskets including for hot-air balloons and crab- and lobster-pots. There is an increasing trend to use willow for coffins. Small boats,

or coracles, made of willow covered with a waterproof material, can still be found in use in Wales. These uses rely on the strength and light weight of the wood. Willow is also a potential biofuel, burning quickly and cleanly. The amount of carbon dioxide produced is less than that absorbed by the tree during growing so willow removes this greenhouse gas from the atmosphere. Locally, a large area of biofuel willow can be seen along the bank of the Granta River in Grantchester on the Pemberton estate. Willow is grown commercially in Sweden for fuel, sometimes combined with what is known as phytoremediation. This involves applying nutrient-rich wastewater or run-off from landfill sites or industrial facilities to the trees. Organic or heavy metal pollutants are taken up but the trees also benefit from nutrients such as nitrogen and phosphorous.[4]

The goat or pussy willows (*Salix caprea*) have showy catkins ('little cats'). Golden catkins from male trees are sometimes used in churches on Palm Sunday. The cottony seeds are used as nest-lining by goldfinches and by penduline tits to make their hanging nests. The wood is used for axe-handles and clothes pegs. Willow flowers, stems, young leaves and stalks are used to make a Bach Flower Remedy for self-pity.

Weeping willows were introduced to Britain in the 18th century. Legend has it that the poet Alexander Pope planted the first cuttings in Britain, at his home in Twickenham. The Latin species name in *Salix babylonica* is thought to originate from the Bible (Psalm 137). An old translation from the Hebrew is "By the rivers of Babylon where we sat down, ...We hanged our harps upon the willows in the midst thereof ...".[5] However, later translations state the trees were poplars consistent with those that grow along the Euphrates.

The use of willow and salicylates medicinally can lead to side-effects. When salicylates were first used as drugs, typically as sodium salicylate, stomach discomfort was reported. In the 1890s, the German pharmaceutical company Bayer started to investigate the use of other salicylate derivatives as more gentle alternatives. A member of the group working on this was Felix Hoffmann, whose father was taking sodium salicylate for rheumatism and experiencing side-effects. Hoffmann and his colleagues synthesised and tested acetyl salicylic acid which became known as aspirin.[6] The name aspirin may have originated from the original Latin name of meadowsweet (*Spiraea ulmaria*).

Felix Hoffmann's next project for Bayer was to synthesise diacetylmorphine which became known as heroin, supposedly because it made patients feel heroic. When patents were voided during World War I, aspirin manufacture grew and it was one of the most widely-used drugs against the Spanish flu pandemic in 1918. As part of the Treaty of Versailles in 1919, Bayer was forced to give up their rights to the trademarks aspirin and heroin. In the post-war period, the use of aspirin diversified and included the combination of soluble aspirin (or disprin) and bicarbonate of soda to produce Alka-Seltzer.

The link between salicylates and bleeding led to their use as anti-clotting drugs. While use for pain relief and as an anti-inflammatory has declined with the introduction of newer drugs such as aminoacetophen (paracetamol) and ibuprofen, the use of a daily low dose of aspirin to prevent heart attacks and strokes has grown. Aspirin has also been shown to be effective against various cancers.[7] Recent estimates suggest a global usage of over 100 billion tablets per year.[8]

In the body, aspirin is thought to reduce production of prostaglandins by inhibiting the enzyme cyclooxygenase (COX). At sites of inflammation, prostaglandins can sensitise nerve endings and increase blood flow causing pain and swelling.[9]

Salicylic acid is used today as an anti-acne treatment, in anti-dandruff shampoo and in over-the-counter treatments for warts and verrucae. The related compound methyl salicylate, also known as oil of wintergreen, is used for

its antiseptic properties in mouthwash and for its rubefacient properties in treating joint and muscle pain. It can also be used in small quantities as a food flavouring.

References

1. J.F. Nunn, (1996); *Ancient Egyptian Medicine*, British Museum Press, London, ISBN 0-7141-1906-7, p 158
2. L. Frost, A. Griffiths, (2001); *Plants of Eden*, Alison Hodge, ISBN 090672029X, p 62
3. E. Stone, (1763); *Philosophical Transactions (1683–1775)*, 53, 195–200, http://www.jstor.org/stable/105721
4. I. Dimitriou, P. Aronsson, (2005); FAO Corporate Document Repository, http://www.fao.org/docrep/008/a0026e/a0026e11.htm
5. http://www.twickenham-museum.org.uk/detail.asp?ContentID=401 [Accessed 05.02.2014]
6. W. Sneader, (2000); *British Medical Journal*, 321, 1591–4, doi: 10.1136/bmj.321.7276.1591
7. e.g. P.M. Rothwell, M. Wilson, C.-E. Elwin, B. Norrving, A. Algra, C.P. Warlow, T.W. Meade, (2010); *The Lancet*, 376(9754), 1741–50, doi: 10.1016/S0140-6736(10)61543-7
8. http://www.aspirin-foundation.com [Accessed 14.10.2013]
9. J.R. Vane (1971); *Nature New Biology*, 231, 232–5, doi: 10.1038/newbio231232a0

January

15

16

17

18

19

20

21

January

22

23

24

25

26

27

28

3. Black Pine (*Pinus nigra*) and the Pinenes

The Old Pinetum in the Cambridge University Botanic Garden dates from its foundation in 1846. The black pines shown here (*Pinus nigra* ssp. *nigra* (left) and ssp. *salzmannii* (right)) were planted opposite each other on the Main Walk by the Garden's founder, John Henslow, to show the extremes of variation in form at the two extremes of distribution. The former example, native to Austria, has only a few branches which grow downwards with short needles, to allow snow to slide off. In the latter example, found in central and southern Spain where snow is not an issue, the branches spread upwards and have large needles.

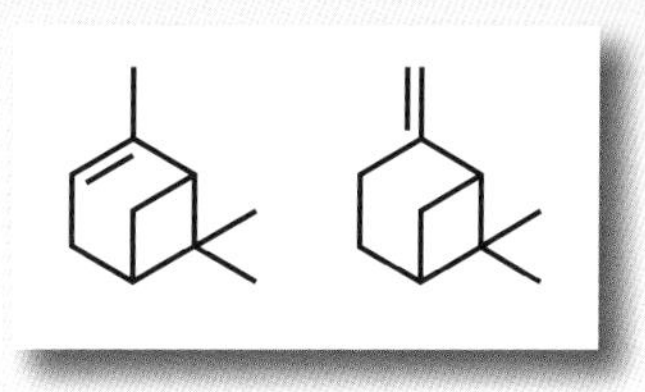

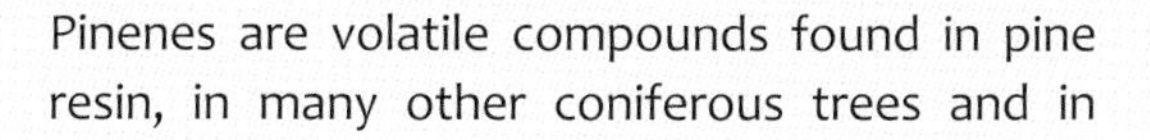

Pinenes are volatile compounds found in pine resin, in many other coniferous trees and in eucalyptus and rosemary oils. They belong to a class of compounds called the terpenes and exist as two isomers, α- and β-pinene (chemical diagrams left and right, respectively), which differ in the position of the double bond. Pinenes are the main constituents of natural turpentine which is used as a solvent and a starting material in the synthesis of many other compounds. The volatile compounds are also used by insects in their chemical communication system.[1]

Pinenes released by the tree react with ozone and other particles in the air, in the presence of sunlight, to form nanometer-sized particles (nanoparticles). Due to their size, these are termed Rayleigh scatterers in that they split white light from the sun into its components, scattering different colours (wavelengths) of light by different amounts. As blue light is scattered most, pine trees can appear to be shrouded in a blue haze. The Blue Ridge Mountains in Virginia, so called because of their characteristic haze, are heavily covered in pine trees. As the quantity of pinenes released depends on the temperature, the mechanism gives the trees natural temperature control. Recent research has shown how the low-volatility aerosol particles that make up the haze are formed. They reflect sunlight and also facilitate cloud formation so it has been suggested they could limit the effects of global warming.[2]

Pine essential oil comes from the steam distillation of stems, twigs and needles. It is deodorising and anti-bacterial and is used as a disinfectant. In aromatherapy, it is used to reduce inflammation, clear mucus, boost the immune system and fight a range of infections. The needles and buds are used similarly in herbal medicine where extracts are used for respiratory infections and as a stimulant to treat arthritis and rheumatism.[3]

Pine essence was used in an inhaler for the treatment of bronchitis known as Hiram Maxim's Pipe of Peace.[4] Sir Hiram Maxim (1840–1916) was an American inventor who was based in Britain after 1881. His most famous invention is the Maxim machine gun which was used extensively by both sides during World War I. He also invented the first automatic sprinkler for buildings and was in dispute with Thomas Edison over his claims to have invented the electric light-bulb. He was fascinated by flight and his fairground ride, the Captive Flying Machine, was used in many amusement parks. One is still in operation today in Blackpool Pleasure Beach. Maxim suffered from bronchitis, aggravated by the London smog, and developed the inhaler to treat his ailment. Dirigo, as the pine essence was termed, was Maxim's own concoction and this was combined with menthol, from mint. Here, both the volatility of the compounds and their anti-bacterial properties were utilised. Inhalation therapy benefits from delivering the treatment right to the affected area (i.e. the lungs). Although Maxim swore by the results, his friends were concerned about the damage this enterprise would do to his reputation. Challenged on his use of 'quackery' Maxim compared this reaction with that to his machine-gun, 'It is a very creditable thing to invent a killing machine but nothing less than a disgrace to invent an apparatus to prevent human suffering'.[5]

When the volatile terpenes are removed from the liquid resin, a solid form known as rosin (or colophany) remains. Rosin contains resin acids such as abietic acid in varying proportions depending on the species. Abietic acid is the primary irritant in pine wood and resin. Rosin is used in printing inks and paper, sealing

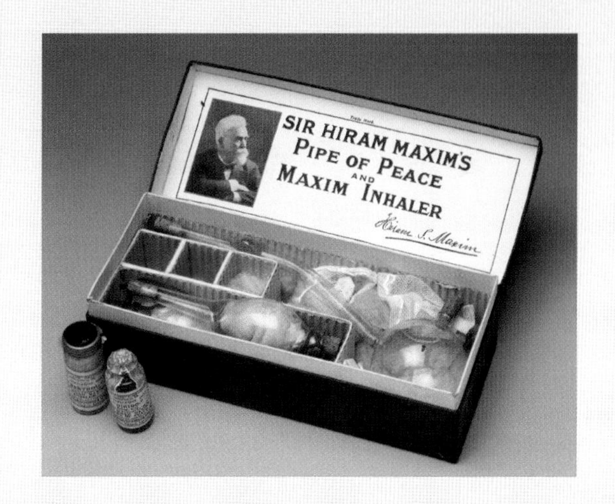

wax, varnishes and glues. It is used in soldering fluxes to improve metal flow. In the food and pharmaceutical industries, rosin is used as a glazing agent (E915) e.g. in chewing gum. Other uses for rosin and its derivatives in the pharmaceutical industry are in ointments, as a binding agent and as a matrix former in tablet formulations. It also acts as a microencapsulating agent, allowing controlled release of the active drugs while it is itself biodegradable.[6] Many other uses of rosin rely on its ability to increase friction. It is rubbed on the bows used to play stringed instruments, such as violins and cellos, so the bow strings grip better and the vibration is clearer. Powdered rosin is sometimes used by gymnasts, dancers, boxers and weightlifters to improve grip or traction. Rosin soaps can be made by reacting rosin with caustic alkalis such as sodium or potassium hydroxide.

Pine wood is used in house joists, rafters, flooring, telegraph poles, chipboard, furniture and shipbuilding. The flowers, stems, young leaves and stalks are used to make a Bach Flower Remedy for feelings of guilt.

High temperature treatment of pine wood in the absence of air produces pine tar. This is used as a wood preservative and sealant. Traditionally, it was used to waterproof ropes made of natural fibres. Handling the treated ropes would stain the hands of sailors leading to the nicknaming of British sailors as tars or Jack Tars. Medicinally, pine tar soap has been used to treat skin conditions such as eczema and psoriasis. Tar is used in the rubber industry as a softener and as an antiseptic, germicidal treatment for horses' hooves.

Pine nuts are edible but only around 20 species produce large enough nuts to warrant commercial harvesting. These include Korean pine (*Pinus koraiensis*), chilgoza pine (*Pinus gerardiana*) and stone pine (*Pinus pinea*). The nuts are used in a wide range of sweet and savoury dishes e.g. the Italian savoury sauce pesto. Nuts can also be used to make pine nut coffee or piñón, popular in south west USA. The name of the character Pinocchio, in the book by Carlo Collodi and made into a film by Disney, is from the Tuscan for pine nut or pine eye. Consumption of nuts can lead to so-called 'pine-mouth' in some people. This is a taste disturbance which develops 12–48 hours after consumption and can last 1–2 weeks. Although it is thought to be harmless, it decreases appetite and the enjoyment of food. The condition may be linked to nuts from a specific variety of pine tree but the causes are still unknown.[7] One suggestion is that it may be due to the chemicals used in the shelling process.

References

1. S.J. Seybold, D.P.W. Huber, J.C. Lee, A.D. Graves, J. Bohlmann, (2006); *Phytochemistry Reviews*, 5, 143–78, doi: 10.1007/s11101-006-9002-8
2. M. McGrath, (2014); http://www.bbc.co.uk/news/science-environment-26340038 [Accessed 03.03.2014]
3. D. Hoffmann (1996); *Complete Illustrated Guide to the Holistic Herbal*, Element Books, ISBN 0-00-713301-4, p 124
4. M. Sanders, (2007); *Primary Care Respiratory Journal*, 16(2), 71–81, doi: 10.3132/pcrj.2007.00017
5. http://www.s104638357.websitehome.co.uk/html/birthday_maxim_04.htm [Accessed 14.10.13]
6. S. Kumar, S.K. Gupta, (2013); *Polimery w Medycynie*, 43(1), 45–8, http://www.polimery.am.wroc.pl/pdf/191.pdf
7. G. Möller, (2010); *Food Technology*, 64(5), http://www.ift.org/food-technology/past-issues/2010/may/features/online-exclusive-pine-nuts.aspx

January

29

30

31

February

1

2

3

4

February

5

6

7

8

9

10

11

4. Carrots (*Daucus carota*) and β-Carotene

β-Carotene is a red-orange pigment found in many plants including carrots. It can be used as a food colouring when it is known as E160a. Although carotenes get their name from carrots, β-carotene is actually more abundant in sweet potato than in carrots. Other good sources are kale, spinach and butternut squash.[1] β-Carotene is a conjugated polyene – i.e. it has a chain of alternate double and single carbon-carbon bonds. This causes it to be sensitive to sunlight but also to mop up free radicals (which have a 'spare' electron) in the body. It reacts readily to give a series of related compounds, the carotenoids.

Many animals get their characteristic colour from eating foods containing carotenes and carotenoids including flamingos, shrimps and salmon.

Wild flamingos obtain carotenes from shrimps which in turn, obtain carotenes from algae. Flamingos living in captivity are given food enriched with carotenes to maintain their colour. A study of greater flamingos shows that they spread carotenoid pigments mixed with wax, excreted from glands near their tails, over their feathers.[2] The colour is believed to make them more attractive to potential mates and the behaviour is more frequent among female flamingos. As the pigments are photosensitive and fade with time, they need to be re-applied frequently.

Over-consumption of carotenes can also cause an orange colouration in humans, particularly on the palms of the hands and the soles of the feet. This harmless condition is known as carotenodermia and has been linked to drinking carrot juice or orange drinks containing E160a as a colouring agent. β-Carotene is a precursor of vitamin A (retinol) so, although carrots don't specifically aid night vision, they are linked to good eye health. This may be predicted by the Doctrine of Signatures as a carrot slice shows characteristic radiating lines resembling those in the pupil and iris of the eye.

Carotenes protect the cardiovascular system from damage from oxidation. One study[3] looked at the protection afforded by fruit and vegetable consumption categorised by the colour of the fruit and vegetables. Orange/yellow coloured fruit and vegetables were found to be the most effective and within this group, carrots were found to be the single most risk-reducing food.

In carrots, the β-Carotene occurs in the tap roots which, in Western carrots, are commonly orange. There has been speculation that the orange colouration was the result of selective breeding by Dutch growers in the 16th century, to honour William of Orange. However, there is no documentary evidence for this and, in fact, the House of Orange was not favoured by the Dutch government at this time. The aim of selective breeding may have been to make the carrots less bitter rather than to change their colour.

As well as orange, carrots come in a range of colours from white, cream and yellow to purple and black. Eastern or Asiatic carrots originated in Central Asia and are typically purple. Although purple carrots are purple on the outside, they are orange on the inside, and contain as much β-Carotene as orange carrots do. The purple colour comes from a different class of pigments known as anthocyanins which also have antioxidant properties and are responsible for the autumnal colouration of leaves. On cooking purple carrots, the anthocyanins dissolve in the cooking water turning it brown and can stain cookware. The purple colour was once used to dye the royal robes of the ancient Afghans and is used today to colour candles and foods.

Carrots have been cultivated for over 2000 years but wild carrots (*Daucus carota*) have been around even longer. Seeds have been found estimated to be over 10,000 years old. Wild carrots are indigenous to Europe, North Africa and

parts of Asia. They have a less well-developed root than the cultivated varieties grown as a food crop. Their characteristic clusters of small white flowers are referred to in one common name, Queen Anne's Lace. The feathery leaves were popular in ladies' head-dresses in the reign of James I and VI.[4]

Wild carrot seeds were used as an aphrodisiac in ancient Rome possibly related to their oestrogen content. In some cultures, they were used as a method of contraception. Later medicinal uses include for the treatment of asthma, gout, rheumatism, cystitis, kidney stones, as a diuretic and for the treatment of colic and flatulence.

Other current or suggested uses of carrots are as an ingredient of biofuels and the seed oil as an industrial lubricant.[5] A new material, Curran, made from 80% carrot fibre and 20% carbon fibre has been manufactured into fishing rods and used to fabricate a Formula 3 racing car steering wheel. Wild carrots are rich in vitamin A and are sometimes used in anti-wrinkle creams.

Other beneficial compounds in carrots include falcarinol which prevents fungal infection of the plant. It also has anti-cancer, anti-bacterial and anti-inflammatory activities. Falcarinol is a polyacetylene, active in the reduced form. The anti-oxidant properties of β-Carotene support this by preventing oxidation and these two classes of phytonutrients work together to maximise the health benefits. The anti-oxidant luteolin found in carrots has been linked to a reduction in age-related memory loss[6] and also shows some anti-cancer properties.

As well as the roots, carrot greens or tops also contain many useful nutrients. They are in the same botanical family as hemlock (*Conium maculatum*) and are sometimes treated with suspicion. However, they contain more vitamin C than the root and also potassium and calcium. Potassium can lower blood pressure and help prevent osteoporosis by reducing calcium loss. Their green colour comes from chlorophyll, a source of magnesium which regulates blood pressure and promotes strong bones and muscles. The greens, unlike the roots, are a good source of vitamin K which is good for bone health. They have anti-cancer and antiseptic properties and can be juiced and used as a mouthwash.

Carrots are eaten all over the world with China the biggest producer. There are numerous varieties, at least one common variety for every letter of the alphabet![7] Unfortunately, the popularity of carrots has led to development of pest-resistant, faster growing varieties which have caused the nutrient content to successively diminish with each generation.[8]

They are among the most popular of all vegetables and because of their relative sweetness, are palatable to children. Youngsters crave fatty and sugary food and reject bitter tasting foods (like Brussels sprouts) as potential poisons! The bitterness is amplified by the number of taste-buds children have – around 30,000 when babies but only a third of this by the time they reach adulthood. The preference for sweet foods does not start to diminish until puberty, and sweet tasting vegetables like carrots and sweetcorn are often firm favourites at least until then.

References

1. U.S. Department of Agriculture, Agricultural Research Service, (2013). USDA National Nutrient Database for Standard Reference, Release 26. Nutrient Data Laboratory Home Page, http://www.ars.usda.gov/ba/bhnrc/ndl
2. J.A. Amat, M.A. Rendón, J. Garrido-Fernández, A. Garrido, M. Rendón-Martos, A. Pérez-Gálvez, (2011); *Behavioural Ecology and Sociobiology*, 65(4), 665–73, doi: 10.1007/s00265-010-1068-z
3. L.M. Oude Griep, W.M.M. Verschuren, D. Kromhout, M.C. Ocké, J.M. Geleijnse , (2011); *British Journal of Nutrition*, 106(10), 1562–9, doi: 10.1017/S0007114511001942
4. Mrs M. Grieve, (1973 edition); *A Modern Herbal*, Merchant Book Company Ltd., ISBN 1904779018, p 161
5. J. Stolarczyk, J. Janick, (2011); *Chronica Horticulturae*, 51(2), 13–20, http://www.actahort.org/chronica/pdf/ch5102.pdf
6. S. Jang, R.N. Dilger, R.W. Johnson, (2010); *Journal of Nutrition*, 140(10), 1892–8, doi: 10.3945/jn.110.123273
7. http://www.carrotmuseum.co.uk/atoz.html [Accessed 15.10.13]
8. D.R. Davis, M.D. Epp, H.D. Riordan, (2004), *Journal of the American College of Nutrition*, 23(6), 669–82, doi: 10.1080/07315724.2004.10719409

February

12

13

14

15

16

17

18

February

19

20

21

22

23

24

25

5. Lupins (*Lupinus* spp.) and Sparteine

Lupinus is a genus of over 200 species which produce tall spikes of thickly clustered flowers in a variety of colours. Each flower has a wide upper petal termed a 'flag' and two fused lower petals termed a 'keel', which are symmetrical along the vertical axis. *Lupinus* is Latin for wolf-like or belonging to a wolf and its use for the genus may be connected to the plants' reputation as prolific killers of livestock. A number of species found in south west USA are known as bluebonnets (e.g. *Lupinus texensis*, Texas bluebonnet) and these are the state flowers of Texas. The flowers are thought to resemble the shape of the bonnets worn by pioneer women.

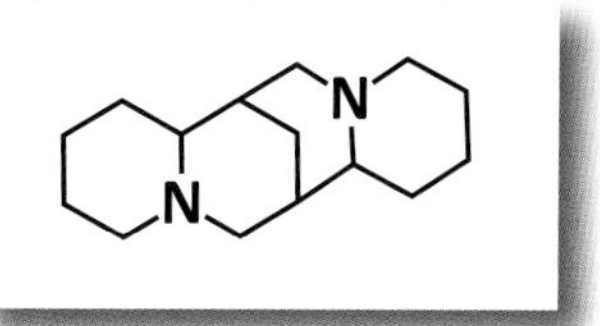

Lupins can fix nitrogen from the atmosphere, rendering the soil fertile for other plants and this, along with their deep root system, means lupins can grow in poor quality soil. They were described as the best plants for light soil such as on the Suffolk coast.[1] After the eruption of Mount St. Helens in 1980 (Washington State,

USA), *Lupinus lepidus* plants were the first to grow on the barren landscape. They can be grown as companion plants, increasing the nitrogen in the soil for other plants. They are an important larval food source for some butterflies and moths e.g. the mission blue butterfly.

Lupins contain toxic, bitter tasting alkaloids primarily sparteine and lupinine, to deter grazers. Sparteine is the most abundant alkaloid in *Lupinus mutabilis* and is a class 1a anti-arrhythmic agent (sodium channel blocker). Sparteine is an oxytocic – it facilitates childbirth by stimulating contractions of the uterus. It is a chiral base used as a starting material in organic synthesis.

Although lupin beans (seeds) are toxic due to their alkaloid content, they are an attractive potential food source as they contain all of the essential amino acids as well as protein and fat and can act as a substitute for soya beans. They have more protein and less fat than soya beans, have anti-oxidant properties and are also considered to be a prebiotic. They can be rendered safe by careful preparation and are sometimes used as a foodstuff in the Mediterranean Basin and Latin America. Low alkaloid varieties (termed sweet lupins) have also been developed and these are commonly used as a food for livestock. They can also be milled into a flour or fashioned as a drink, lupin milk, and yellow lupins are used as an egg substitute. Lupin flour can be mixed with wheatflour (with lupin contents 5–15%) for its nutritional and food-processing benefits and until recently was often a 'hidden' ingredient in foods. Due to their emulsifying properties, lupin extracts have potential uses in the meat industry.

However, allergy to lupins can be a significant problem particularly for individuals with a peanut allergy. This may be related to the α-conglutin protein found in lupin beans. Since 2006, the presence of lupin-based ingredients has been required to be indicated on food packaging in the EU.[2]

The α-conglutin protein has, however, been linked to beneficial effects. It can cross the intestinal barrier and reduce blood glucose levels. This could account for the anti-diabetic properties of lupin beans.[3] Another compound from yellow lupin (*Lupinus arabicus*) beans, termed 'calcium-anhydro-oxydiamine phosphate' or Magolan was suggested for the treatment of diabetes in the early 20th century. The recom-

mended dose was 3 grains several times a day.[4] The bruised seeds (beans) of white lupin (*Lupinus albus*) were sometimes applied externally to treat ulcers and internally as a diuretic, anthelmintic (to destroy worms) and emmenagogue (to promote menstruation or prevent pregnancy). Other traditional uses of lupins were to make soaps, to provide a fibre to make paper and in adhesive.[1]

The use of lupin (or lupini) beans was common among the Romans and their use spread throughout the Roman Empire. They were used as pieces of money by Roman actors leading to the saying 'nummus lupinus' meaning a spurious piece of money.[1] *Lupinus mutabilis* or chocho, the Andean variety of lupin, was used as a food within the Incan empire. Lupin beans are often sold in brine in jars and eaten as a snack food. The process required to render lupin beans safe for consumption depends on the alkaloid content. For high-alkaloid varieties, the beans are soaked overnight in water, boiled in salted water for two hours, placed in running water for seven to fourteen days, boiled again in salted water and finally pickled, if they are not to be canned! In Italy, the traditional method for the step where running water is required, is to soak the beans in a pillowcase or fabric bag in a stream.

Newly-bred varieties of sweet lupin contain only safe levels of alkaloids and require no soaking at all. These are grown extensively in Germany. As they can be grown in temperate climates, they are becoming a viable cash crop alternative to soya. Lupins are also grown in Australia, Europe, Russia and the Americas as a green manure,

livestock fodder and high protein additive for human and animal foods. If crops of new, sweet lupins are cross-pollinated with wild, bitter lupins this could render the crop toxic and precautions have to be taken to avoid this.[5]

There are numerous possible symptoms of poisoning from ingestion of incorrectly prepared beans. These include confusion, dilated pupils, disorientation, fever, raised blood pressure and heart rate, tremors, stomach pain, dizziness and anxiety. Lupin bean alkaloids are anticholinergic, affecting the peripheral and central nervous systems. The severity of the symptoms depends on the quantity of the alkaloids consumed but in severe cases lupin poisoning can be fatal.

Another issue with using lupins for livestock is what is known as lupinosis. This is caused by toxins produced by the fungus *Diaporthe toxica* which grows within lupin stems. Livestock grazing on lupin stubble will be exposed and may develop the disease, characterised by liver damage. Sheep are more susceptible than cattle but by careful management the effects can be minimised. Feeding the livestock lupin beans before allowing them to eat the stubble encourages them to search for the beans within the stubble so they consume less of the toxin.[6]

References

1. Mrs M. Grieve, (1973 edition); *A Modern Herbal*, Merchant Book Company Ltd., ISBN 1904779018, pp 502–3
2. U. Jappe, S. Vieths, (2010); *Molecular Nutrition and Food Research*, 54(1), 113–26, doi: 10.1002/mnfr.200900365
3. J. Capraro, A. Clemente, L.A. Rubio, C. Magni, A. Scarafoni, M. Duranti, (2011); *Food Chemistry*, 125(4), 1279–83, doi: 10.1016/j.foodchem.2010.10.073
4. J.P. Remington, H.C. Wood, Jr., (1918), *Dispensatory of the United States of America*, J. B. Lippincott, Philadelphia; scanned version online at http://www.henriettesherbal.com/eclectic/usdisp/index.html
5. M. Richards, (2010); www.dpi.nsw.gov.au/__data/assets/pdf_file/0008/186713/testing-albus-lupins-for-bitter-seed.pdf
6. http://archive.agric.wa.gov.au/OBJTWR/imported_assets/content/fcp/lp/lup/lupins/Lupinbulletinch13.pdf

February

26

27

28/29

March

1

2

3

4

March

5

6

7

8

9

10

11

6. Sweet Woodruff (*Galium odoratum*) and Coumarin

Sweet woodruff (*Galium odoratum*) contains the compound coumarin as do sweet clover (*Melilotus* spp.) and cinnamon (*Cinnamomum* spp.). The name coumarin derives from cumaru, the common name of the flowering tree *Dipteryx odorata*, which produces tonka beans, another rich source. Coumarin is a sweet smelling compound, recognisable as the smell of newly mown hay. The smell intensifies on drying the plant leading to its traditional use in pot pourri, to fill mattresses, ward off moths and scent laundry. Dried leaves are sometimes used in high quality snuffs. Coumarin is widely used in the perfume industry, forming the basis of the fougere family of scents and is an ingredient in around 90% of perfumes. The fixative properties along with the notes of vanilla, almond and liquorice add to its popularity.

Coumarin is bitter tasting, moderately toxic, has appetite-suppressant properties and is present in the plants to deter grazers. It can be used medicinally to treat lymphoedema[1]

and although it has no anti-coagulant properties itself, it can be transformed synthetically or by some fungi into dicoumerol, which is an anti-coagulant (i.e. stops blood clotting). Dicoumerol is the cause of sweet clover disease which was observed in cattle grazing on mouldy hay from a field containing some sweet clover.[2]

This discovery led scientists in Wisconsin to try to develop related anti-clotting drugs and they successfully synthesised the drug warfarin (**W**isconsin **A**lumni **R**esearch **F**oundation-**arin**) from 4-hydroxycoumarin. This was approved for use in 1954 and remains the most popular oral anti-coagulant drug. However, there are some issues with its use – it interacts with foods high in vitamin K1, such as leafy vegetables. The action of the molecule in the body is to deplete active vitamin K1 and this vitamin can be used as an antidote in cases of warfarin overdose. Warfarin does not thin the blood but prevents the formation of clotting factors (which require the presence of vitamin K1). The dose has to be carefully monitored by blood testing to ensure the risk of bleeding is not excessive while maintaining a high enough quantity to give a therapeutic benefit.[3] An early recipient of warfarin was US president Dwight D. Eisenhower, who was prescribed the drug after a heart attack in 1955. There has been a suggestion that Nikita Khrushchev and others used warfarin to poison Josef Stalin.[4]

Another important use of warfarin is as rat poison. It has been used as a rodenticide since

1948 and as it is odourless and tasteless, it can be mixed with food bait. Rodents will, therefore, return over a period of days to feed on the bait until the lethal dose of warfarin has been accumulated. Warfarin's use as a rat poison has started to decline due to the introduction of other effective poisons such as the 4-hydroxycoumarin derivatives brodifacoum, a 'super-warfarin' and coumatetralyl and due to some rodent populations becoming resistant.

The presence of coumarin in cinnamon can have implications for human health. Coumarin is moderately toxic to the liver and kidneys but although banned from use as a food additive in many countries, it is often present in food or drink either naturally e.g. in chamomile tea and bison grass vodka (Żubróka) or as an additive in some tobacco products, for example. There are two major types of cinnamon – Chinese cinnamon (*Cinnamomum cassia*) which contains a relatively large amount of coumarin and is typically used in foodstuffs and Sri Lankan cinnamon (*Cinnamomum verum*) which contains virtually none.

In 2006 there was a food scare connected to the use of cinnamon in traditional German star-shaped Christmas biscuits. Due to the sensitivity of some individuals to coumarin, the German

food safety agency (the Federal Institute for Risk Assessment) recommended consumption of no more than 0.1 milligram per kilogram body weight per day. This is termed a tolerable daily intake, TDI, which can safely be consumed over a lifetime with no ill-effects, even for sensitive individuals. In 2012, updated recommendations, considering bioavailability data, stated that with the coumarin content of star biscuits at the highest permitted level, the TDI equated to infants consuming 6 star-biscuits a day and adults 30 a day assuming no other sources of coumarin in the diet.[5]

Sweet woodruff is traditionally used to flavour May Wine (or Maibowle), an aromatised wine drunk on May Day in Germany and parts of Belgium. Medicinal uses include as a treatment for insomnia, liver disease and kidney stones, to strengthen the heart, as an anti-inflammatory, diuretic and anti-spasmodic. Topically, it can be used as a compress to treat varicose veins, phlebitis and ulcers and was traditionally applied to wounds to aid healing. In medieval times, soldiers sometimes carried sprigs of woodruff in their helmets to bring success in battle.[6] A red dye can be made from the roots while a light brown dye is produced from the stems and leaves.

The plant grows to under a foot tall and produces small white flowers in May and June. Leaves appear in whorls from the stalks just below the flowers. In spite of its diminutive size, sweet woodruff is also known as master of the woods. Ripe seeds are covered with hooked bristles and are spread by passing animals or birds to which they attach. The related species cleavers (*Galium aparine*) reflects its similar behaviour in its name (cleave meaning to latch on). The genus name comes from the Greek word gala for milk, related to the use of lady's bedstraw flowers (*Galium verum*) to curdle milk in cheese-making. This species is also known as cheese rennet but other members of the genus share this property. The stem and leaves of lady's bedstraw give a yellow dye while the roots produce a red colour. The dyes were used to colour cheese and fabric.

References

1. N. Farinola, N. Piller, (2005); *Lymphatic Research and Biology*, 3(2), 81–6, doi:10.1089/lrb.2005.3.81
2. M.A. Stahmann, C.F. Huebner, K.P. Link, (1941); *Journal of Biological Chemistry*, 138(2), 513–27
3. M.D. Freedman, (1992); *Journal of Clinical Pharmacology*, 32(3), 196–209, doi: 10.1002/j.1552-4604.1992.tb03827.x
4. J. Brent, V.P. Naumov, (2003); *Stalin's last crime: the plot against the Jewish doctors, 1948–1953*, Harper Collins Publishers, New York, ISBN: 0-06-019524-X, p 322
5. http://www.bfr.bund.de/en/press_information/2012/26/cassia_cinnamon_with_high_coumarin_contents_to_be_consumed_in_moderation-131836.html [Accessed 16.10.2013]
6. T. Breverton, (2011); *Breverton's Complete Herbal*, Quercus Publishing Ltd, ISBN 978-0-85738-336-5, pp 324–5

March

13

14

15

16

17

18

19

March

20

21

22

23

24

25

26

7. Garlic (*Allium sativum*) and DADS

Garlic (*Allium sativum*) contains alliin and the enzyme allinase in separate compartments. When the cell walls are ruptured by cutting or chewing these compounds come into contact, reacting to produce allicin. This is unstable and rapidly produces a range of sulfur containing compounds including diallyl disulfide (DADS), a yellow liquid, one of the principal components of distilled garlic oil. Other compounds include diallyl trisulfide and diallyl tetrasulfide.

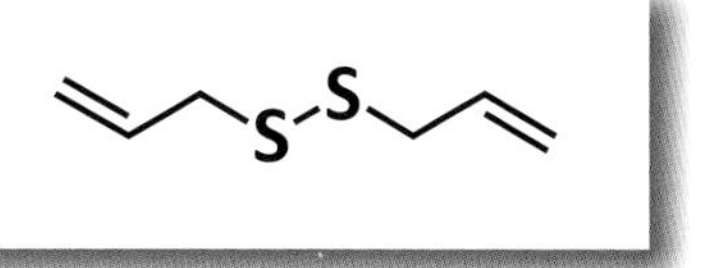

The production of allicin forms the plant's protection against attack. It is active for very short periods, localised at the site of the cell wall breakage, killing invaders but decomposing before there is significant damage to the plant. Allicin has very good anti-bacterial, anti-fungal and anti-cancer properties. However, its instability provides a challenge for scientists looking to harness these properties for medicinal use. Production of stabilised allicin is now possible and this has been shown to be effective against MRSA,[1] multi-resistant cystic fibrosis, amoebic dysentery, pneumonia, hay fever, and to both

prevent infection and reduce duration of the common cold and flu.

The sulfurous compounds give garlic its characteristic penetrating smell. It is so diffusive that if garlic is rubbed on the soles of the feet, the odour is exhaled by the lungs. The odour is perceived by activation of the Transient Receptor Potential Cation Channel Member A1 (TRPA1). This channel is present in humans, other animals and some fungi and is a sensor for environmental irritants, pain, cold and stench. Sensory nerve endings are irritated causing the lachrymatory effect of garlic and other related plants such as onion. It is likely this channel developed to protect the plant from predators at an early stage of evolution. It is also activated by the structurally similar allyl isothiocyanate, the pungent component of wasabi and other mustard plants in the genus *Brassica*.[2]

DADS has anti-microbial, fungicidal and insecticidal properties and is present in preparations for the selective decontamination of organs before surgery.[3] It acts against the stomach ulcer germ, *Heliobacter pyroli*, although less effectively than allicin.[4] It also has anti-cancer and cardioprotective properties but it is the principal cause of skin reaction to garlic (allergic contact dermatitis). The allergy usually starts in the fingertips and is not prevented by wearing gloves as DADS penetrates most commercially available gloves.[5]

DADS can also be used as an environmentally friendly nematicide and is used in the food industry to improve the taste of meat, vegetables and fruit. It acts to detoxify the cells in the body, preventing inflammation and this is linked to many of its beneficial effects and those of garlic. The effect may also be related to the observation that garlic increases the potency of some medicines such as non-steroidal anti-inflammatory drugs and antibiotics.[6]

Garlic is used as a food flavouring and taken as a supplement for high blood pressure and to improve heart health. The activation of the TRPA1 receptors by DADS and allicin causes vasodilation and this leads to a short-term reduction in blood pressure.[2] Garlic also thins the blood by reducing the sticking together of platelets. Garlic has been reported to prevent colorectal cancer and DADS may be a significant contributor to this.[7]

In herbal medicine, garlic oil is used to treat respiratory infections, whooping cough and, with other herbs, bronchitic asthma.[8] Garlic is believed to support the development of natural bacterial flora in the digestive tract while killing pathogenic organisms. It is also used to reduce cholesterol and, externally, to treat ringworm.

Crushed garlic extracts have been used to treat a wide range of infectious diseases including dysentery, typhoid, cholera, smallpox and tuberculosis. The antiseptic properties of garlic were verified by Louis Pasteur in 1858 but it had been used prior to this as an anti-bacterial, anti-fungal, anti-viral and anti-parasitic. During World War I, garlic was in great demand as an antiseptic. In 1916, the British government offered one shilling per pound of garlic bulbs, hoping to acquire tons. The raw juice from these was diluted with water and applied on sterilised swabs of *Sphagnum* moss to prevent suppuration of wounds.[9] In the 1920s, its widespread use declined due to the introduction of the sulfonamide class of antibiotic drugs which also contain sulfur. One of garlic's common names, Russian penicillin, refers to its use as an alternative to other antibiotics by the Red Army during World War II.[10]

Garlic was considered to be an aphrodisiac and was forbidden to monks. Traditionally, a person running a race might chew a piece of garlic in the belief it would prevent competitors from getting ahead of them. Similarly, Hungarian jockeys reportedly fastened garlic to the bits of their horses to deter other horses from passing[9] and the Romans ate it before battle to give them strength. Garlic was also believed to protect against vampires. A concoction containing garlic, known as 'Four Thieves' Vinegar', got its name when four thieves confessed to using it to protect themselves from disease while they robbed plague victims' bodies. Garlic was also used for protection during the plague outbreak in Marseilles in 1722.[9]

Garlic varieties are classified as hardneck or softneck. Unlike softneck garlics, hardneck varieties usually form a stalk (or scape) at the top of which grows the 'flower'. This is not actually a flower but a collection of miniature bulbs known as bulbils. The scapes are removed if garlic is being grown for bulbs, as scape and bulbil growth diverts resources away for the developing bulbs. The removed 'flowers' can be eaten either in salads or, if a little older, lightly cooked. The production of long curling scapes results in some hardneck varieties being nicknamed 'serpent garlic'. Garlic is in the same genus as onions (*Allium cepa*) and chives (*Allium schoenoprasm*), both of which contain characteristic sulfurous compounds and are used to add flavour to food.

Chives (*Allium schoenoprasm*)

References

1. N.J. Bennett, P.D. Josling, R. Cutler, (30.10.08); *European Journal for Nutraceutical Research*; access provided by the authors
2. D.M. Bautista, P. Movahed, A. Hinman, H.E. Axelsson, O. Sterner, E.D. Högestätt, D. Julius, S.-E. Jordt, P.M. Zygmunt, (2005); *Proceedings of the National Academy of Sciences*, 102(34), 12248–52, doi: 10.1073/pnas.0505356102
3. J. Yu, Y.B. Xiao, X.Y. Wang, (2007); *Chinese Journal of Traumatology*, 10(3), 131–7, PubMed PMID: 17535634
4. E.A. O'Gara, D.J. Hill, D.J. Maslin, (2000); *Applied and Environmental Microbiology*, 66(5), 2269–73, doi:10.1128/AEM.66.5.2269-2273.2000
5. M. Moyle, K. Frowen, R. Nixon, (2004); *Australasian Journal of Dermatology*, 45(4), 223–5, doi: 10.1111/j.1440-0960.2004.00102.x
6. T. Breverton, (2011); *Breverton's Complete Herbal*, Quercus Publishing Ltd, ISBN 978-0-85738-336-5, pp 158–9
7. H.J. Jo, J.D. Song, K.M. Kim, Y.H. Cho, K.H. Kim, Y.C. Park, (2008); *Oncology Reports*, 19(1), 275–80
8. D. Hoffmann, (1996); *Complete Illustrated Guide to the Holistic Herbal*, Element Books, ISBN 0-00-713301-4
9. Mrs M. Grieve, (1973 edition); *A Modern Herbal*, Merchant Book Company Ltd., ISBN 1904779018, pp 342–5
10. B.B. Petrovska, S. Cekovska, (2010); *Pharmacognosy Reviews*, 4(7), 106–10, doi: 10.4103/0973-7847.65321

March

27

28

29

30

31

April

1

2

April

3

4

5

6

7

8

9

8. Nettles (*Urtica dioica*) and Formic Acid

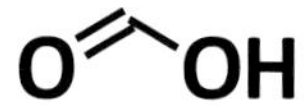

The Latin name for common or stinging nettles, *Urtica dioica*, derives from urere 'to burn' and 'two houses' referring to the separation of male and female flowers on different plants. Stinging nettles have hollow stinging hairs with tips which break off when touched. The hair has a high silica content like glass and it acts as a mini hypodermic syringe, injecting the stinging chemicals stored at the base of the hair under the skin. The sting contains a cocktail of chemicals including formic acid, histamine, moroidin, serotonin and acetylcholine. The sting causes a burning sensation and hives. However, although this is painful and inconvenient, the effect of our native nettle's sting is mild in comparison with that of some of its cousins. For example, the sting of one species in Java, *Urtica urentissima* or Devil's leaf, causes symptoms which can last for weeks, occasionally causing death.

The presence of the sting deters grazers which make nettles an attractive food source for many species of insects and larvae, including red admiral, small tortoiseshell and peacock

butterflies. However, nettles repel flies so a bunch was traditionally hung in the larder.

The sting is released by brushing against the hairs but if the nettle is grasped firmly and the hairs are flattened, no stinging occurs. In one of Aesop's fables 'The Boy and the Nettles', the boy tells his mother "Although it pains me so much, I did but touch it ever so gently." "That was just it" said his mother, "which caused it to sting you. The next time you touch a Nettle, grasp it boldly, and it will be soft as silk to your hand, and not in the least hurt you." The moral of the story is whatever you do, do it with all your might.[1]

Dock (*Rumex* spp.) often grows close to nettles and the leaves are used to treat nettle stings, as suggested by the Doctrine of Signatures. There may be some scientific basis to this. Dock leaves contain a small amount of the anti-histamine chlorphenamine which may act against the histamine contained in the sting. However, they also contain oxalic acid, which does not cause stinging like formic acid but does make the sting feel worse. In addition, rubbing the leaf on the sting may cause the pain to spread further. Other possible treatments for nettle stings include the juice of nettles which contains anti-histamines or calamine lotion. Serotonin is chemically related to melatonin, is associated with feelings of well-being and has an effect on appetite and sleep. However, it is also thought to increase the pain associated with the sting and is present in many plant spines and in insect venom.

Formic acid is the simplest carboxylic acid and is also found in bee and ant stings. The name derives from the Latin word for ant – *formica*. Unlike most animals, anteaters do not have concentrated hydrochloric acid in their stomachs to help digest food. Instead, they rely on the formic acid they get from the ants in their diet. Formic acid was first isolated by the distillation of ants in 1670, by English naturalist John Ray[2] but a synthetic process was developed in the 19th century.

Formic acid is used as a preservative and anti-bacterial agent in livestock feed. It is used in the production of leather, textiles and rubber[3] and in cleaning products such as limescale remover. Formic acid fuel cells have the potential to provide an environmentally friendly, cost-effective alternative to lithium batteries. In the fuel cell,

formic acid and oxygen react to produce hydrogen and carbon dioxide releasing a large amount of energy.[4] These cells could, potentially, be used to power mobile phones and laptop computers. The acid has low toxicity but it is one of the products of the metabolism of methanol along with formaldehyde and has been linked to optic nerve damage and blindness caused by methanol poisoning.

Nettles have been used for a variety of medicinal and other purposes for generations. Native American braves used to flog themselves with nettles (urtication) to help them stay awake while on watch. This was also a traditional remedy for rheumatism used by Roman soldiers posted to Britain.

Nettles are a good source of calcium, magnesium and iron as well as a range of vitamins and other trace elements. The young leaves taste like spinach and, once the stinging chemicals have been removed by soaking or cooking, can be made into soup or added to stews. Nettle tea is still enjoyed as a spring tonic and nettles can also be made into a beer. Drying the leaves also removes the sting and dried nettles are sometimes added to livestock feed. The seeds can be fed to horses to improve the condition of their coats. Similarly, nettle shampoo makes hair glossy and has anti-dandruff properties. Nettles are used in cheese-making in the production of the Cornish cheese, Yarg. This is wrapped in fresh nettle leaves before it is matured and the leaves form an edible rind. Nettles are also used as a flavouring for some varieties of Gouda and as a vegetarian source of rennet.

Mature nettle stems can be woven into a cloth (nettlecloth) which was traditionally used to make tablecloths and sheets in Scandinavia and Scotland. A Bronze Age body was found in Denmark wrapped in nettlecloth possibly from Austria, suggesting trading between these countries was taking place in 800 BC.[5] During World War I when Britain controlled 90% of the world's cotton crops, Germany developed nettlecloth as a fabric for making soldiers' uniforms and rucksacks, with uniforms being made of up to 85% nettlecloth. More recently, a mixed nettlecloth and wool fabric has been used for upholstery which benefits from the nettles' natural fire-retardant properties. Ötzi, the Iceman, a corpse found near the Austrian/Italian border in 1991, frozen since about 3,200 BC, was carrying arrows. The fletchings of these, used to give them stability during flight, were attached with birch tar then wound with nettle fibres.[6] The presence of nettles in a location is a good indicator to archaeologists that the area was formerly a human settlement. Nettles thrive in nitrogen-rich soils and grow where there has been an abundance of human waste products.

Nettles also provide a green dye from the stems and leaves and a yellow dye on boiling the roots. During World War II, nettles were harvested by British schoolchildren to produce the green dye for camouflage. This is currently used as a food colouring for canned vegetables in Germany. The potential of nettles as an environmentally friendly crop is helped by their abundance and ability to grow without the use of pesticides.

In herbal medicine, nettles are credited with strengthening and supporting the whole body. They are used to treat childhood eczema, as an astringent to stop nose bleeds and haemorrhaging[7] and as a galactagogue (to promote lactation). Nettle leaves prevent inflammation therefore can be used to treat arthritis and gout, while nettle root has proven activity against the symptoms of benign prostatic hyperplasia.[8] The latter effect may be due to the presence in the root of 3,4-divanillyltetrahydrofuran which can occupy sex-hormone binding globulin, increasing the level of free testosterone.[9] Extracts of nettle root are sometimes taken by bodybuilders for this reason. Dried leaves are a natural anti-histamine and also have anti-asthmatic properties. Nettle plants are covered in small hairs therefore their uses to treat respiratory conditions and to promote hair growth are suggested by the Doctrine of Signatures. Nettles are a good source of boron which may account for their success in treating rheumatoid[10] and osteoarthritis.

References

1. http://www.litscape.com/author/Aesop/The_Boy_and_the_Nettles.html
2. J. Wray, (1670); *Philosophical Transactions*, 5(57–8), 2063–6, doi: 10.1098/rstl.1670.0052
3. W. Reutemann, H. Kieczka, (2011); *Ullmann's Encyclopedia of Industrial Chemistry*, doi: 10.1002/14356007.a12_013.pub2
4. Y. Zhu, S.Y. Ha, R.I. Masel, (2004); *Journal of Power Sources*, 130(1–2), 8–14, doi: 10.1016/j.jpowsour.2003.11.051
5. http://news.ku.dk/all_news/2012/2012.9/nettles_reveal_long_distance_bronze_age_trade_connections/
6. A. Fleckinger, (2011); *Ötzi, the Iceman The Full Facts at a Glance*, 3rd updated edition, Folio, Vienna/Bolzano and the South Tyrol Museum of Archaeology, Bolzano, ISBN 978-3-85256-574-3, p 81
7. D. Hoffmann, (1996); *Complete Illustrated Guide to the Holistic Herbal*, Element Books, ISBN 0-00-713301-4, p 158
8. M.R. Safarinejad, (2005); *Journal of Herbal Pharmacotherapy*, 5(4), 1–11, doi: 10.1080/J157v05n04_01
9. M. Schöttner, D. Ganßer, G. Spiteller, (1997); *Planta Medica*, 63(6), 529–32, doi: 10.1055/s-2006-957756
10. Z.S. Al-Rawi, F.I. Gorial, W.A. Al-Shammary, F. Muhsin, A.S. Al-Naaimi, S. Kareem, (2013); *Journal of Experimental and Integrative Medicine*, 3(1), 9–15, doi: 10.5455/jeim.101112.or.053

April

10

11

12

13

14

15

16

April

17

18

19

20

21

22

23

9. Hops (*Humulus lupulus*) and Isoadhumulone

Hops grow on vines (bines) which tend to twist and curl around things, gripping with small, stiff hairs. The Latin name of the species common hop, *Humulus lupulus*, refers to its wolf-like nature – the vines bind tightly and do not readily let go. The word hop comes from the Anglo-Saxon *hoppan*, to climb. They are closely related to cannabis. The earliest recorded use of hops was 4,000 years ago to preserve yeast for brewing and baking. They were also traditionally used in dyeing as the stems produce a brownish-red dye while the flowers and leaves produce a yellow one. The stems can be pulped for paper-making or soaked and beaten to yield fibres that can be used to make rope or a coarse cloth. The Romans ate hop shoots the way we eat asparagus.[1]

Like in nettles, male and female flowers occur on different plants. The female flower heads, cones, are used in beer making and are made up of bracts and bracteoles surrounding the cone. Underneath the bracteoles are the lupulin glands containing the resin and oils. Today, 98% of all hops grown are used in beer-making. Only female hops are grown in commercial hop fields and these reproduce vegetatively from stem cuttings. Hops can also be found growing in hedgerows.

Hops were fermented into a drink *symthum* by the ancient Egyptians and their use in brewing

can be traced to medieval Europe. In the 12th century, a distinction was made between beer and ale, the former containing hops and the latter containing none. Hops were initially imported from Europe and were not grown in Britain until the early 16th century. There was resistance in Britain to the adulteration of ale with foreign hops and Henry VIII described hops as "a wicked weed that would spoil the taste of the drink and endanger the people."[2] Some cities like Coventry and Norwich even banned their use by brewers.

However, there were practical reasons why the use of hops prevailed. Ale has a limited shelf life and would often turn sour, particularly in hot weather. On sea voyages, the water was undrinkable so beer was carried for the sailors (about a gallon per sailor per day). The heavily hopped India Pale Ale (IPA) became popular on long sea journeys as it did not spoil. As beer was commonly drunk in Europe, areas of Britain with large immigrant populations, such as the Dutch weavers in East Anglia, had a big demand for beer. By the 16th century use of hops was common.[3]

Harvesting hops was often carried out by migrant populations along with city families who travelled to hop-growing areas like Kent for the harvest. An annual working holiday to the hop fields was a regular feature of the lives of thousands of working class families from London's East End as they supplemented the travelling Romany working population. Families worked together – men sometimes standing on stilts, cutting down the vines which were gathered below by women and children and the cones separated for drying. This continued into the 1950s, by which time mechanical hop pickers had replaced the labour-intensive harvesting system. Earlier in the annual hop-growing cycle, the Romanies were essential to the springtime training or 'twiddling' of the new shoots up their supporting wires. Working with hops was not without problems – around 3% of people are allergic and contact results in skin lesions.

The most common variety of hops in Britain is Fuggles, introduced by Richard Fuggle around 1875 and now used in 90% of British beer. The word fuggled is used to describe a person having consumed too much of the resulting beer!

Hops are added towards the end of the brewing process after drying in kilns known as oast houses and the mixture of wort (the unfermented sweet liquid infusion from ground malt) and hops boiled for 1½ hours. During this time, humulones in the hops are converted to isohumulones. These compounds sterilise the beer and make it bitter. Isohumulone is bacteriostatic towards gram positive bacteria and, therefore, protects the beer from serious contamination by this type of bacteria.

Humulone (α-lupulic acid, α-bitter acid) is an α-acid and converts on boiling to a mixture of cis- and trans-isohumulone. It has anti-oxidant, COX-2 inhibitor, anti-viral and anti-bacterial properties.[4] Other significant α-acids in hops are adhumulone and cohumulone. These are also isomerised on boiling to isoadhumulone and isocohumulone.

The bitterness of beer is quoted in International Bitterness Units (IBUs), where 1 IBU is equivalent to 1 part per million (ppm) isohumulones. These values are measured by passing light through an acidified beer sample. The amount of light absorbed at 275 nm is proportionate to the quantity of isohumulones present. Most normal strength lagers are about 10 IBUs, Guinness is about 40 IBUs and IPAs are above 40 IBUs, with some well over 100.

Isohumulones are light-sensitive and if beer is exposed to light, they decompose producing free radicals. These react with sulfur-containing amino acids to produce thiols which give beer a skunky flavour.[5] To protect against this, beer bottles are often brown but as these are more expensive than clear bottles, some brewers keep bottles boxed to exclude light or use modified hop products. Adding a slice of lime when serving can also be used to mask the odour.

Varieties of hop with high α-acid content and a low essential oil content produce beers with a high degree of bitterness – these are known as bittering hops. The essential oils add flavour to the beer and where this is more important, different varieties, known as finishing hops, with a higher essential oil content, are used. The oils are readily removed by boiling so finishing hops are added later in the process than bittering hops. The most important compounds in the oils include humulene, the main component of so-called noble hops and myrcene which gives citrus and pine aromas to the beer. Hops also contain β-acids which, like α-acids, are bitter tasting. However, they are less soluble and do not isomerise to iso-acids. Their effect on taste increases with time as they oxidise. The β-acids are also important in preserving the beer and tests on hops β-acids (HBAs) have shown they

can be used to prevent listeria contamination of foods.[6]

Hops have a long history of use in herbal medicine as a sedative. Pillows containing hops and lavender are used to promote sleep and a combination of hops and valerian can be taken orally to aid sleep. The sedative properties of hops are believed to be related to compounds including 2-methyl-3-buten-2-ol which modulate receptors in the brain and inhibit the central nervous system.[7] However, they should be avoided by people suffering from depression as hops may accentuate this.[8] Other traditional uses include to treat liver and digestive complaints and, externally, to treat ulcers. Strangely, they also, reputedly, reduce the desire for alcohol. Some recent research suggests a possible benefit in the treatment of menopausal symptoms.[9] This is due to the presence of one of the most oestrogenic of the phyto-oestrogens known, 8-prenylnaringenin and means hops could, potentially, be used in hormone replacement therapy. It may also be related to the use of hops in treating hair loss. According to the Doctrine of Signatures, this is also suggested by the hairiness of the plant, particularly the stems.

Hops are also used to flavour drinks other than beer – they are an ingredient of the Swedish soft-drink, Julmust and the Latin American soft-drink, Malta. While hops are widely used by humans, they are toxic to dogs causing panting, high body temperature, seizures and sometimes death.

References

1. L. Frost, A. Griffiths, (2001); *Plants of Eden*, Alison Hodge, ISBN: 090672029X, p 48
2. Mrs M. Grieve, (1973 edition); *A Modern Herbal*, Merchant Book Company Ltd., ISBN 1904779018, pp 411–2
3. B. Laws, (2010); *Fifty Plants that Changed the Course of History*, David and Charles, ISBN: 0-7153-3854-4, pp 110–5
4. N. Yamaguchi, K. Satoh-Yamaguchi, M. Ono, (2009); *Phytomedicine*, 16(4), 369–76, doi: 10.1016/j.phymed.2008.12.021
5. C.S. Burns, A. Heyerick, D. De Keukeleire, M.D.E. Forbes, (2001); *Chemistry – A European Journal*, 7(21), 4553–61, doi: 10.1002/1521-3765(20011105)7:21<4553::AID-CHEM4553>3.0.CO;2-0
6. C. Shen, I. Geornaras, P.A. Kendall, J.N. Sofos, (2009); *Journal of Food Protection*, 72(4), 702–6
7. L. Franco, C. Sánchez, R. Bravo, A. Rodriguez, C. Barriga, J.C. Cubero, (2012); *Acta Physiologica Hungarica*, 99(2), 133–9 doi: 10.1556/APhysiol.99.2012.2.6
8. D. Hoffmann, (1996); *Complete Illustrated Guide to the Holistic Herbal*, Element Books, ISBN 0-00-713301-4, p 102
9. A.M. Keiler, O. Zierau, G. Kretzschmar, (2013); *Planta Medica*; 79(07), 576–9, doi: 10.1055/s-0032-1328330

April

24

25

26

27

28

29

30

May

1

2

3

4

5

6

7

10. Spurges (*Euphorbia* spp.) and Ingenol Mebutate

The genus *Euphorbia* consists of over 2,000 diverse species, sometimes known as spurges. The name spurge originates from the Middle English/Old French word *espurge* or *espurgier*, meaning to purge, relating to the purgative properties of the sap which exudes from the wounds when the plant is cut or damaged. The Latin name *euphorbia* can be traced back to Pliny who used it in AD 79. He was schooled in botany by Juba II who was treated by the Greek physician Euphorbus using a medicinal plant, believed to be what is now known as resin spurge (*Euphorbia resinifera*). This cactus-like plant is native to Morocco. In Greek, the word *euphorbus* means well-fed and it has been suggested Juba chose the name euphorbia as both the plant and his physician were a little fleshy. Juba sent an expedition to the Canary Islands and more species of *Euphorbia* were found, one of which was named in his honour, *Euphorbia regis-jubae*. Juba's first wife was Selene, the daughter of Cleopatra and Mark Antony.[1]

The very small flowers are unisexual and aggregated into clusters known as cyathia. This latter feature is unique in the plant kingdom. Variation in the structure and appearance of the cyathium results in the wide diversity of species seen. One

of the most common is poinsettia (*Euphorbia pulcherrima*) of which there are hundreds of cultivars. This is the most widely-grown pot-plant in Europe at Christmas. The Latin name translates as 'most beautiful Euphorbia'. Poinsettia was named after Joel R. Poinsett, an American physician and diplomat, who introduced the plant to the USA from Mexico in 1825.

The sap of spurges congeals on contact with air and gums up the mouths of predators trying to graze on the plant. The skin irritant effects are mainly caused by di- and tri-terpenes such as betulin and related compounds called esters and terpenoids. Betulin is also found in the bark of birch trees. The sap is irritating to mucous membranes (eyes, nose and mouth) and the effects worsen if it is not washed off immediately. Sap from some species can cause blindness therefore eye protection is necessary when working with these plants.[2] When large succulent spurges are cut in an enclosed space like a greenhouse, precautions need to be taken to avoid breathing the vapour as it can cause irritation to the eyes and throat. The sap of poinsettia is phototoxic, becoming more of an irritant in sunlight.[3]

There are numerous uses for the sap (or latex) of species of *Euphorbia*. In Africa, Transvaal candelabra tree (*Euphorbia cooperi*) sap is drained from the branches and boiled, then the resin painted on tree branches to catch birds. Sections of the branches or tufts of grass soaked in latex are also thrown into water to stun fish, allowing them to be caught.[4] *Euphorbia hyberna* (sometimes *hiberna*) (Irish spurge) was used for the same purpose by peasants in Ireland. A small basket containing bruised leaves was placed in the river and, reputedly, stupefied fish for

several miles down-stream.[5] At the concentrations found in the fish, the latex is not toxic to humans and the fish can be safely eaten.

Mixed with sugar, the latex of *Euphorbia ingens* (candelabra tree) is used as a laxative and to treat alcohol dependency but it is extremely poisonous and overuse can be fatal. The latex is believed to be a cure for ulcers and some cancers. The light wood of this tree-like species is sometimes used to make boats.[4] Before the wood is cut, a fire is made at the foot of the tree. The heat causes the sap to coagulate and the branches and trunk can be cut without danger of contact with the sap. The *Euphorbia* trees produce a multitude of small flowers which attract bees. However, the honey produced has the same burning effect as the sap and is commonly known as 'noors honey'. The burning effect is intensified on drinking water.

Cypress spurge (*Euphorbia cyparissias*) was used in France as a powerful purgative and was also known as *rhubarbe des pauvres* (poor man's rhubarb).[5] The latex of the tree species *Euphorbia tirucalli* (pencil tree, sticks on fire) was used to repel flying insects and as an arrow poison. In addition, it was believed to cure infertility, cancers, warts and some sexually transmitted diseases. Rice boiled with the latex was used to

catch birds for food. Extracts have been used as a substitute for petroleum and to produce a low grade rubber.[4]

Boiling the leaves or stems of *Euphorbia antisyphilitica* with dilute sulfuric acid produces candelilla wax. It was originally used to make candles hence the name 'little candle'. During World War I, the wax was used to waterproof tents and other military equipment. It can be used as a vegan alternative to beeswax, as a glazing agent or binder in food (E902) and to make varnish. In the cosmetics industry, it is used as a component of lip balm and in moisturisers. As its name suggests, *Euphorbia antisyphilitica* was traditionally used as a treatment for syphilis, as was *Euphorbia hirta.*

Pill-bearing spurge or asthma plant (*Euphorbia hirta*) is, as you might expect from its name, widely used medicinally. A preparation of the aerial parts of the plant has a relaxing effect on the smooth muscles in the lungs. Thus, the plant is used to treat bronchitis, spasms in the larynx and upper respiratory catarrh. It has also been suggested as a treatment for *Plasmodium* infections[6] and dengue fever. It is used in Ayurvedic medicine (traditional Indian medicine) for digestive problems, tumours, dysentery, jaundice and 'female disorders' as well as for respiratory complaints. A review of its proven medicinal benefits identifies anti-bacterial, anti-inflammatory, anti-cancer, anti-oxidant, anti-fertility, anti-asthmatic, anti-amoebic and anti-fungal effects and more.[7]

Native to India, *Euphorbia caducifolia* latex was traditionally used to aid wound-healing and to treat skin eruptions and other skin diseases. Recent research has confirmed that it

does have significant wound-healing properties.[8] Peking spurge (*Euphorbia pekinensis*) is one of the 50 fundamental herbs in traditional Chinese medicine, where it is called dà jǐ. The root is used to treat oedema, kidney problems, tuberculosis and epilepsy, to relieve swelling and to disperse nodules.

An extract from the latex of *Euphorbia peplus* (petty spurge) has recently been identified as a treatment for actinic keratosis, a common skin condition triggered by over-exposure to the sun. If left untreated, the characteristic sandpaper patches can develop into squamous cell carcinoma, a non-melanoma form of skin cancer.[9] The active ingredient is ingenol mebutate and a gel formulation of the drug has been licensed in the USA and Europe for treatment of this condition under the trade name Picato. Ingenol mebutate has also been shown to be effective against other human non-melanoma skin cancers.[10] It is alternatively known by the common name ingenol angelate i.e. it is related to angelic acid. This acid was first isolated from the plant angelica (*Angelica archangelica*). *Euphorbia peplus* is commonly known as radium weed or cancer weed in Australia where it is used as a traditional skin cancer remedy.

References

1. http://www.euphorbiaceae.org/pages/about_euphorbia.html [Accessed 5.10.2013]
2. T. Eke, S. Al-Husainy, M.K. Raynor, (2000); *Archives of Ophthalmology*, 118(1), 13–16, doi: 10.1001/archopht.118.1.13
3. A. Massmanian, (1998); *Contact Dermatitis*, 38(2), 113–4, doi: 10.1111/j.1600-0536.1998.tb05668.x
4. S. Gildenhuys, (2006); *Euphorbia World*, 2(1), 9–14, http://www.euphorbia-international.org/journal/pdf_files/EW2-1-sample.pdf
5. Mrs M. Grieve, (1973 edition); *A Modern Herbal*, Merchant Book Company Ltd., ISBN 1904779018 p 765
6. L. Tona, R.K. Cimanga, K. Mesia, C.T. Musuamba, T. De Bruyne, S. Apers, N. Hernans, S. Van Miert, L. Pieters, J. Totté, A.J. Vlietinck, (2004); *Journal of Ethnopharmacology*, 93(1), 27–32, doi: 10.1016/j.jep.2004.02.022
7. S. Kumar, R. Malhotra, D. Kumar, (2010); *Pharmacognosy Review*, 4(7), 58–61, doi: 10.4103/0973-7847.65327
8. M. Goyal, B.P. Nagori, D. Sasmal, (2012); *Journal of Ethnopharmacology*, 144(3), 786–90, doi: 10.1016/j.jep.2012.10.006
9. M. Lebwohl, N. Swanson, L.L. Anderson, A. Melgaard, Z. Xu, B. Berman, (2012); *New England Journal of Medicine*, 366(11), 1010–9, doi: 10.1056/NEJMoa1111170
10. J.R. Ramsay, A. Suhrbier, J.H. Aylward, S. Ogbourne, S.-J. Cozzi, M.G. Poulsen, K.C. Baumann, P. Welburn, G.L. Redlich, P.G. Parsons, (2011); *British Journal of Dermatology*, 164(3), 633–6, doi: 10.1111/j.1365-2133.2010.10184.x

May

8

9

10

11

12

13

14

May

15

16

17

18

19

20

21

11. The Mints (*Mentha* spp.) and Menthol

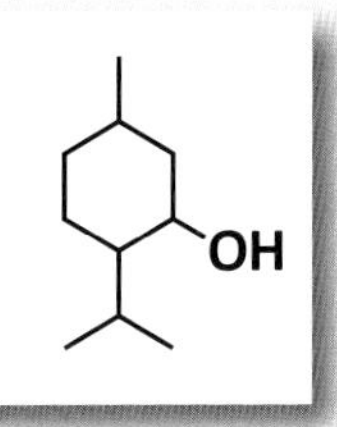

The genus *Mentha* contains several species which have characteristic smelling oils and square stems. Due to their tendency to spread unchecked, they are often considered invasive and grown in pots. Planted near roses, mints can deter aphids. They have a wide range of uses from culinary and medicinal to cosmetic. Mint leaves are often used as a food flavouring and mint sauce is commonly served with lamb. Mint has carminative properties and lamb is often a fatty meat therefore mint sauce can help with digestion as well as improving the flavour.

Mint sauce is made using spearmint (*Mentha spicata*) and this mint is one of the ingredients of the Cuban cocktail mojito and the American drink mint julep. It was traditionally used to prevent milk curdling. A spearmint cultivar, *Mentha spicata* Nana, is used to flavour Touareg tea, popular in Arabian countries. Spearmint is also used as a flavouring for toothpaste, chewing gum and confectionery and to add fragrance to shampoos and soaps.

Mojito

The chemical components of spearmint oil include (R)-(-)-carvone, a chiral terpene which is responsible for the characteristic smell. The mirror image of this compound, (S)-(+)-carvone, is found in caraway and dill seeds. (R)-(-)-carvone is used directly as a flavouring for chewing gum, as an air-freshener and as a mosquito repellent. Other compounds include the hydrogenated form, dihydrocarvone, limonene and 1,8-cineol. The latter is also known as eucalyptol and can be extracted from eucalyptus oil.

Spearmint tea has anti-androgenic properties, reducing testosterone levels and is a potential treatment for hirsutism in women.[1] It is an antioxidant and when added to radiation-treated lamb, it was found to delay the oxidation of fats and reduce the formation of harmful products.[2]

Spearmint contains almost no menthol but peppermint (*Mentha x piperita*), corn mint (*Mentha arvensis*) and horse mint (*Mentha longifolia*) all contain significant quantities. Peppermint is a hybrid of spearmint and water mint (*Mentha aquatica*). It is widely cultivated and the leaves and flowering tops harvested. Peppermint oil is among the most popular essential oils and English peppermint is grown commercially near Wisbech in Cambridgeshire, among other locations. As well as the alcohol menthol, it contains the dehydrogenated compound menthone and menthyl esters such as menthyl acetate. Menthol can be synthesised from thymol obtained from thyme (*Thymus vulgaris*) which is also related to the mint family.[3] Peppermint has been shown to protect against radiation.[4]

Menthol or peppermint produce a cooling sensation by activating the transient receptor potential cation channel subfamily M member 8 (TRPM8).[5] This receptor is also known as the cold and menthol receptor 1 (CMR1). Although the cold sensors are activated, menthol does not actually lower the body temperature. The compound capsaicin, from chilli peppers, behaves similarly towards the heat sensors, activating them chemically without any change in temperature.

The activation of cold receptors in the nose causes respiration to slow down and in

combination with the 'cooling' effect, breathing feels easier. As a result, menthol is an ingredient in many lozenges and rubs (e.g. Vicks VapoRub) used for cold symptoms. Along with pine essence, menthol was used in Hiram Maxim's Inhaler for the treatment of bronchitis. Higher concentrations of menthol can be an irritant and also have local anaesthetic effects. This has led to the use of creams containing menthol, like Tiger Balm, for topical pain relief. The analgesic properties of menthol are due to selective activation of κ-opioid receptors.[6]

Menthol is used in lip balms, to treat sunburn and itching, in shaving products, as a food flavouring and in oral hygiene products. The latter use is related not only to the fresh smell of menthol and mint oils but also to their anti-bacterial and anti-microbial properties. In perfumery, it is used to enhance the floral notes of rose, geranium and lavender. It can be used to relax smooth muscles and as an anti-spasmodic during upper gastrointestinal endoscopy[7] and helps treat nausea and motion-sickness. Menthol is used as a treatment for tracheal mites in honey bees.[5]

Mentholated cigarettes are popular in some cultures. They mask the smoky smell when smoked but menthol inhibits the metabolism of nicotine thus increasing systemic exposure.[8]

The idea that peppermint improves concentration and memory was tested by pumping the vapour into a classroom in a Liverpool school and testing pupils' performance. Anecdotal evidence suggested an improvement and this was backed up by recent reports from a study at Wheeling Jesuit University. During a seven week experiment, run by the Psychology Department, the performance of students was correlated with their use of peppermint scented pencils (smencils). Students using the smencils most often had the best grades while students who didn't use them at all had the worst grades.[9]

Peppermint can be used to deter rats, mice, moths and ants and, medicinally, the oil is used to treat irritable bowel syndrome (IBS). Peppermint oil is a natural insecticide. The compound pulegone which is also present in horse mint (*Mentha longifolia*) and pennyroyal (*Mentha pulegium*), is the most important contributor to this. Pulegone can be metabolised to the highly toxic menthofuran which has been linked to death in some cases where pennyroyal was self-administered to bring about abortion. Pennyroyal essential oil should never, therefore, be taken internally.

Thyme (*Thymus vulgaris*)

Pennyroyal or Corsican mint (*Mentha requienii*) are used as companion plants, grown near plants of the *Brassica* genus (e.g. cauliflower, broccoli and cabbage). Corsican mint, at only 1–2 inches tall, is the smallest of the mints. It was said to have reached Ireland on a shipwrecked vessel from the Spanish Armada. This mint also provides the flavour for the drink Crème de Menthe. In Greek mythology, Menthe was a water nymph pursued by Hades who was turned into a plant by Hades' wife Persephone. Hades was powerless but he did alter the plant so that when trodden on it would release a beautiful fragrance.

Horse mint (*Mentha longifolia*) is one of four plants used in a preparation for weight-loss in traditional Greco-Arabic and Islamic medicine. A clinical trial showed significant weight-loss and reduction of body mass index (BMI) associated with use.[10]

References

1. P. Grant, (2010); *Phytotherapy Research*, 24(2), 186–8, doi: 10.1002/ptr.2900
2. S.R. Kanatt, R.Chander, A. Sharma, (2007); *Food Chemistry*, 100(2), 451–8, doi: 10.1016/j.foodchem.2005.09.066
3. A.L. Barney, H.B. Hass, (1944); *Industrial and Engineering Chemistry*, 1944, 36(1), pp 85–7 doi: 10.1021/ie50409a018
4. M.S. Balinga, S. Rao, (2010); *Journal of Cancer Research and Therapeutics*, 6(3), 255–62 doi: 10.4103/0973-1482.73336
5. R. Eccles, (1994); *Journal of Pharmacy and Pharmacology*, 46(8), 618–30, doi: 10.1111/j.2042-7158.1994.tb03871.x
6. N. Galeotti, L. Di Cesare Mannelli, G. Mazzanti, A. Bartolini, C. Ghelardini, (2002); *Neuroscience Letters*, 322(3), 145–8, doi: 10.1016/S0304-3940(01)02527-7
7. N. Hiki, M. Kaminishi, T. Hasunuma, M. Nakamura, S. Nomura, N. Yahagi, H. Tajiri, H. Suzuki, (2011); *Clinical Pharmacology & Therapeutics*, 90(2), 221–8, doi: 10.1038/clpt.2011.110
8. N.L. Benowitz, B. Herrera, P. Jacob III, (2004); *The Journal of Pharmacology and Experimental Therapeutics*, 310(3), 1208–15, doi: 10.1124/jpet.104.066902
9. http://www.wtov9.com/news/features/top-stories/stories/research-students-use-scented-pencils-have-higher-gpas-2727.shtml [Accessed 18.02.2014]
10. O. Said, K. Khalil, S. Fulder, Y. Marie, E. Kassis, B. Saad, (2010); *The Open Complementary Medicine Journal*, 2, 1–6, doi: 10.2174/1876391X01002010001

May

22

23

24

25

26

27

28

May

29

30

31

June

1

2

3

4

12. Chilli Pepper (*Capsicum annuum*) and Capsaicin

The genus *Capsicum* (chilli peppers) belongs to the Solanaceae family, one of the most important food families which includes potatoes, tomatoes and aubergines. Members of the genus are best known for their use as a food flavouring, giving 'heat' to a wide range of dishes. They were introduced into Europe in the 16th century to replace black pepper (*Piper nigrum*) which had become difficult to obtain following the fall of Constantinople (Istanbul) to the Ottoman Empire in 1453. The chemical compound responsible for the heat is capsaicin. This is found abundantly in the white pith around the seeds and acts as a deterrent to herbivores. In general, the smaller the chilli the hotter it is as it contains proportionately more seeds. Peppers also contain a number of related compounds, the capsaicinoids e.g. dihydrocapsaicin and homocapsaicin which also contribute to the hotness.

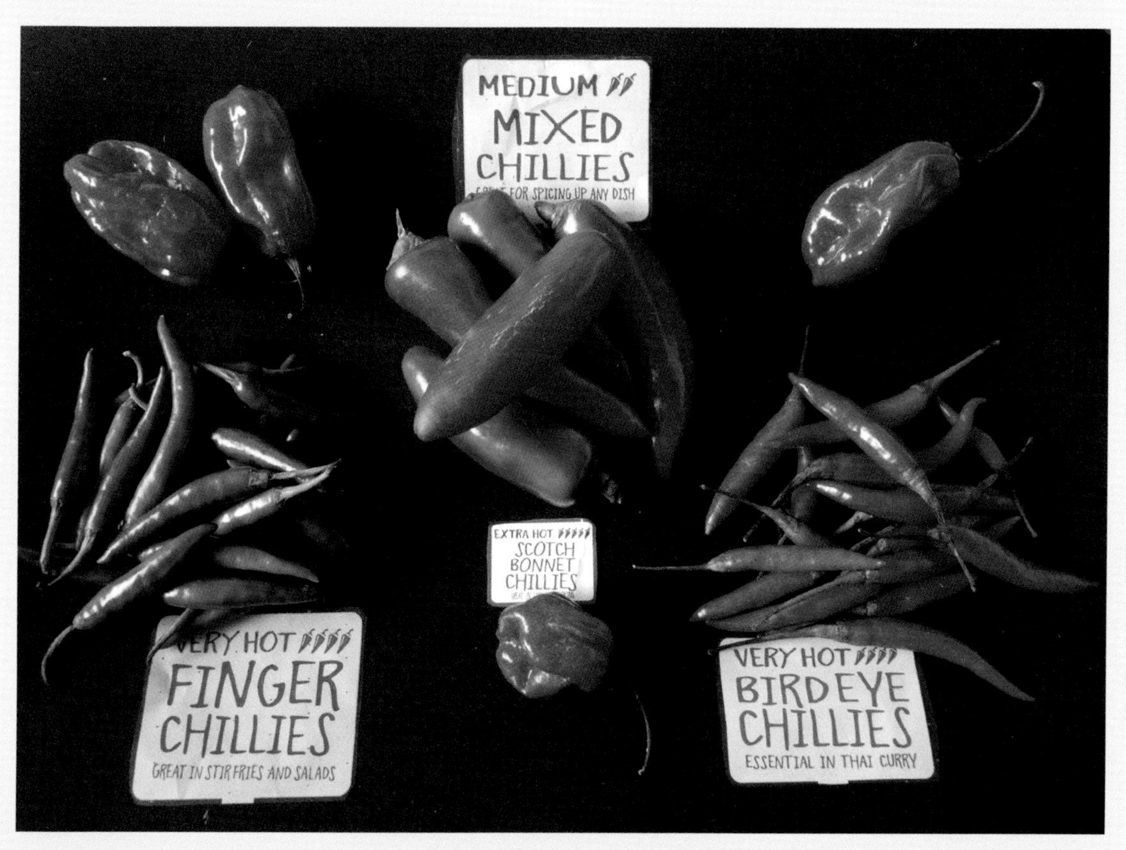

The feeling of heat and burning on consumption of capsaicin is due to activation of the heat sensor transient receptor potential cation channel subfamily V member 1 (TRPV1), often known as the capsaicin receptor. Similarly to the case of menthol with the cold receptor TRPM8, capsaicin activates the heat receptor chemically without raising the temperature of the body. The capsaicin receptor is also activated by the compound allyl isothiocyanate, found in mustard oil and wasabi. Activation causes burning and pain but also an increase in sweating and release of endorphins. The mucous membranes in contact with capsaicin are irritated. Prolonged exposure to capsaicin reduces the sensitivity of the receptors and this may be the root of its analgesic properties. Nerves are overwhelmed and temporarily are unable to report pain. Capsaicin is used in topical ointments to treat pain and inflammation e.g. to treat peripheral neuropathy pain and shingles and to treat psoriasis.

Repeated nasal application of capsaicin has been shown to reduce the occurrence of cluster headaches.[1] Long-term consumption promotes vasorelaxation (widening of blood vessels) and reduces blood pressure.[2] Capsaicinoids reduce the total cholesterol in the bloodstream.[3] Capsaicin has been shown to be active against prostate cancer[4] and can be used as a nasal spray to treat allergies. It is used in pepper spray for personal protection and in law enforcement. As it has pain-relieving and hyper-sensitising properties, it is considered a performance enhancing drug, banned from use in equestrian sport. Four horses entered for the show-jumping event at the 2008 Olympic Games tested positive for capsaicin and were disqualified.[5]

The seeds of *Capsicum* plants are dispersed mainly by birds which do not have the TRPV1 receptor and the seeds pass though their digestive tract intact and germinate later. Mammals consuming the seeds would destroy them by chewing the seeds therefore capsaicin acts as a protection from this. The pain pathway activated by capsaicin is the same one activated by tarantula venom. This was the first proven example of a shared pathway for protection from mammals occurring in both the plant and animal kingdoms.[6]

The best treatment for the burning effect of capsaicin is cold milk. It contains casein which has a detergent effect on capsaicin (i.e. it combines with the capsaicin and negates the burning effect) although a sugar solution is also effective. The Indian side-dish raita, containing cucumber and yoghurt is also used.

Different varieties of chilli contain differing amounts of capsaicin and this is measured to estimate the hotness of the chilli on the Scoville scale. This was named after Wilbur Scoville who developed the Scoville Organoleptic Test in 1912. As originally devised, it involved diluting a solution of the chilli in sugar water until the heat was no longer detectable by a panel of tasters. The degree of dilution gives a measure on the Scoville scale. On this scale, pure capsaicin has a value of 16,000,000 Scoville Heat Units (SHUs), pepper spray 500,000–5,000,000 SHUs, Scotch bonnet (*Capsicum chinense*) 100,000–350,000 SHUs, jalapeños 2,500–5,000 SHUs and bell peppers (a cultivar of *Capsicum annuum*) 0 SHUs. Bell peppers (sweet peppers) are the only peppers not to contain any capsaicin. Today, measurements of capsaicin content are carried out using High Performance Liquid Chromatography (HPLC). This technique allows separation of a mixture into its component parts and determines their quantities, giving more consistent and accurate results. The officially recognised hottest chilli, according to the Guinness Book of Records, is Smokin' Ed's 'Carolina Reaper', grown by the PuckerButt Pepper Company (USA). These peppers had an average value of 1,569,300 SHUs during 2012.[7]

Tabasco sauce is made from tabasco peppers (*Capsicum frutescens* var. *tabasco*). This hot and spicy sauce is widely used as a seasoning and flavouring, famously in the cocktail Bloody Mary. The sauce has a heat of 2,500–5,000 SHUs. The fruits of *Capsicum frutescens* grow upright rather than hanging down from the stem.

The species *Capsicum annuum* has many different varieties of peppers which differ in size, shape and heat. The spice paprika is made from

the ground, dried fruits of either sweet or chilli peppers or a mixture of both. Paprika is used to colour and flavour a range of dishes such as goulash, soups, stews and meats. It is commonly used in Hungary and in several other countries including Morocco, Spain, Turkey and Greece. It can also be used with henna to give a reddish tint to hair. The colour of paprika comes from the carotenoid pigment zeaxanthin found also in goji berries and saffron and used as a food additive E161h. Capsanthin and capsorubin also contribute and are known as E160c when used as a food colouring.

One hybrid, bhut jolokia (or ghost chilli) which has a Scoville rating of over 1,000,000 SHUs, has been identified for possible use in hand grenades by the Indian Army.[8] These could be used in riot control and against terrorists. Other possible uses are as a food additive for use when troops are operating in cold climates and spread on fences around barracks to protect against animals.

Cayenne pepper is a cultivar of *Capsicum annuum* and is also known as Guinea spice or Guinea pepper, named after the city of Cayenne, the capital of French Guiana. It is used when cooking spicy dishes and has a heat of 30,000–50,000 SHUs. In herbal medicine, it is a systemic stimulant used to improve circulation and regulate heart rate and blood flow. Other uses include as a carminative to treat flatulent dyspepsia and colic and applied to the skin as a rubifacient, causing dilation of the capillaries and reduction of pain. It can also be used as an ointment to treat unbroken chilblains.[9]

References

1. B.M. Fusco, S. Marabini, C.A. Maggi, G. Fiore, P. Geppetti, (1994); *Pain*, 59(3), 321–5
2. D. Yang, Z. Luo, S. Ma, W.T. Wong, L. Ma, J. Zhong, H. He, Z. Zhao, T. Cao, Z. Yan, D. Liu, W.J. Arendshorst, Y. Huang, M. Tepel, Z. Zhu, (2010); *Cell Metabolism*, 12(2), 130–41, doi: 10.1016/j.cmet.2010.05.015
3. Y.T. Liang, X.-Y. Tian, J.N. Chen, C. Peng, K.Y. Ma, Y. Zuo, R. Jiao, Y. Lu, Y. Huang, Z.-Y. Chen, (2013); *European Journal of Nutrition*, 52(1), 379–88. doi: 10.1007/s00394-012-0344-2
4. A. Mori, S. Lehmann, J. O'Kelly, T. Kumagai, J.C. Desmond, M. Pervan, W.H. McBride, M. Kizaki, H.P. Koeffler, (2006); *Cancer Research*, 66, 3222–9, doi: 10.1158/0008-5472.CAN-05-0087
5. http://news.bbc.co.uk/sport1/hi/olympics/equestrian/7574220.stm [Accessed 12.11.2013]
6. J. Siemens, S. Zhou, R. Piskorowski, T. Nikai, E.A. Lumpkin, A.I. Basbaum, D. King, D. Julius, (2006); *Nature*, 444, 208–12, doi: 10.1038/nature05285
7. http://www.guinnessworldrecords.com/world-records/1/hottest-chili [Accessed 19.02.2014]
8. http://news.bbc.co.uk/1/hi/8119591.stm [Accessed 12.11.2013]
9. D. Hoffmann (1996); *Complete Illustrated Guide to the Holistic Herbal*, Element Books, ISBN 0-00-713301-4, p 72

June

5

6

7

8

9

10

11

June

12

13

14

15

16

17

18

13. Henbane (*Hyoscyamus niger*) and Scopolamine

Hyoscyamus niger (henbane, stinking nightshade or black henbane) is, like *Capsicum*, from the Solanaceae family which also contains deadly nightshade (*Atropa belladonna*) and mandrake (*Mandragora officinarum*). Mandrake, deadly nightshade and henbane all contain the compound scopolamine, sometimes known as hyoscine, a hallucinogenic compound classified as a tropane alkaloid.

Henbane is toxic and has a long history of misuse. It is one of several plants suggested as being the 'hebona' or 'hebenon' used to kill Hamlet's father in Shakespeare's play. Other suggestions are hemlock or yew.[1] Insane root, a plant eaten to cause madness in medieval times, is usually identified as either henbane or hemlock. Scopolamine was believed to be the poison Dr. (Hawley) Crippen, an American homoeopathic doctor living in London, used to murder his wife, Cora, in 1910. Crippen was

hanged for the crime but, recently, questions have been raised as to the identity of the body parts found. DNA samples suggest the body was male and not related to surviving members of Cora's family. There are no other reported uses of scopolamine in cases of poisoning but Crippen did possess a bottle which, he said, was an ingredient of some of his homoeopathic preparations. There was a huge amount of media interest in the case and it is possible that, under pressure to find a culprit, evidence of Crippin's guilt was planted.[2]

The action of scopolamine is to block the muscarinic receptors, depressing the peripheral and central nervous systems and leading to amnesia and fatigue. It is an anti-emetic and scopolamine patches are used to treat motion sickness. Used post-operatively, it prevents nausea and vomiting and it can be used by scuba divers to improve performance.[3] Other medicinal uses are in anaesthesia, to treat brachycardia (low heart rate), as an anti-spasmodic, an anti-depressant and to treat some types of organophosphate poisoning from pesticides. Side effects include dry mouth (leading to its use when this is desirable e.g. during some surgeries), sedation, blurred vision and urine retention.[4]

As an anaesthetic, it was used in combination with opiates to induce 'twilight sleep' for mothers during childbirth from the 1900s to the 1960s. Henbane was used even earlier for this purpose, in so-called soporific sponges, common in the Middle Ages. These sponges were soaked in plant juices, typically a combination of henbane, opium, hemlock and mandrake and placed under the noses of patients. Not only would patients sleep through their operation but also for many hours afterwards, allowing the body recovery time from the trauma of surgery. This was an early example of using a mixture of agents to give a desired effect and also of the use of inhalation therapy.[5] Henbane was used, historically, to treat mania and in an over-the-counter preparation to treat bronchitis and asthma.

Scopolamine is currently used in both conventional medicine and homoeopathy to treat

stomach cramps in children.[6] Other uses of homoeopathic preparations are to treat cramps and pains accompanying epilepsy, disorders of the bladder and digestive system and to treat mental and emotional problems manifesting as paranoia, jealousy, delusions and aggressive outbursts.

Scopolamine's hallucinogenic effects have led to its recreational use but it has serious side-effects including amnesia and death. In Colombia, its use by criminals in robberies and rape is reportedly common. Scopolamine, known locally as burundanga or the Devil's breath, is obtained from the native borrachero tree (from the genus *Brugmansia*). Victims appear normal but are unable to exert free will, helping robbers take their possessions or withdrawing money from their accounts from ATMs and giving it to the robbers. When the drug wears off, they have no memory of what has taken place. This is believed to be due to the action of scopolamine on the hippocampus in the brain, preventing memories being formed. It means that, even under hypnosis, victims cannot remember what happened to them. Further, as scopolamine acts on the amygdala, victims become submissive. Historically, extracts of the borrachero tree were used to facilitate communication with ancestors and given to wives and mistresses of deceased lords who then followed a suggestion to climb into their own graves!

In 1922, Richard House, a Dallas obstetrician noted that patients given scopolamine during childbirth answered questions truthfully and candidly and that this could be of use when questioning criminals. A person taking scopolamine is unable to construct a lie. This created interest in its possible use as a truth serum from the CIA and elsewhere. Scopolamine may have been used for this purpose by the Czech Secret Police but the undesirable side-effects, including dry mouth and blurred vision were not considered conducive to successful interrogation and its use declined.[8]

Use of henbane was recorded by the ancient Greeks. Pliny (AD 23–79) mentioned its use to yield oracles by the priestesses of Apollo, calling it *Herba Apollinaris*. It has a long association with use in magic potions. The whole plant is toxic although some species are immune e.g. the larvae of some *Lepidoptera* moths. For humans, however, even smelling the flowers can produce giddiness. It can be safely ingested only in very small quantities and henbane was used as a flavouring for beer before the introduction of hops. In 2008, celebrity

Deadly nightshade (*Atropa belladonna*)

chef Antony Worrall Thompson mistakenly recommended the use of henbane leaves in salads. Thompson confused henbane with the plant fat hen (*Chenopodium album*), the young leaves of which can be eaten like spinach.[9] The seed heads resemble a row of teeth and the plant was used in dentistry as suggested by the Doctrine of Signatures. The seeds are held in a capsule which opens transversely. This is surrounded by a hairy, pitcher-shaped calyx with five prickly lobes. A traditional ornament on the head-dress of Jewish high priests was modelled on the seed vessels.[10]

In addition to scopolamine, henbane, deadly nightshade and mandrake also contain the isomers hyoscyamine (L-atropine) and atropine (DL-hyoscyamine). The three related tropane alkaloids all have similar medicinal effects. Deadly nightshade (*Atropa belladonna*) gets its name from its cosmetic use to dilate the pupils, enhancing the beauty of the eyes. Belladonna is Italian for beautiful lady. Egyptian henbane (*Hyoscyamus muticus*) contains a greater quantity of atropine than black henbane does. It was mentioned in the Ebers papyrus (*ca.* 1500 BC), where it was known as Sakran which translates as 'the drunken'. One use was to smoke Sakran to relieve toothache.

References

1. G.J. Higby, (1993); *Pharmacy in History*, 35(3), 137, http://www.jstor.org/stable/41111539
2. Secrets of the Dead; Public Broadcasting Service (PBS); Executed in Error, (2008); http://www.pbs.org/wnet/secrets/episodes/executed-in-error/198/ [Accessed 13.11.2013]
3. T.H. Williams, A.R. Wilkinson, F.M. Davis, C.M. Frampton, (1988); *Undersea Biomedical Research*, 15(2), 89–98, http://archive.rubicon-foundation.org/2495
4. H.P. Rang, M.M. Dale, J.M. Ritter, P.K. Moore, (2003); *Pharmacology* (5th edition); Churchill Livingstone, ISBN 0443071454, p 147
5. P. Juvin, J.-M. Desmonts, (2000); *Anesthesiology*, 93(1), 265–9
6. B. Müller-Krampe, M. Oberbaum, P. Klein, M. Weiser, (2007); *Pediatrics International*, 49(3), 328–34, doi: 10.1111/j.1442-200X.2007.02382.x
7. Y. Strachan, (2012); Is Scopolamine the world's scariest drug? http://digitaljournal.com/article/324779 [Accessed 14.11.2013]
8. G. Bimmerle, "Truth" Drugs in Interrogation (Released 1993), Central Intelligence Agency (CIA), https://www.cia.gov/library/center-for-the-study-of-intelligence/kent-csi/vol5no2/html/v05i2a09p_0001.htm [Accessed 14.11.2013]
9. A. Dawar, (2008); http://www.theguardian.com/lifeandstyle/2008/aug/04/foodanddrink.foodsafety?gusrc=rss&feed=networkfront [Accessed 14.11.2013]
10. Mrs M. Grieve, (1973 edition); *A Modern Herbal*, Merchant Book Company Ltd., ISBN 1904779018 p 400

June

19

20

21

22

23

24

25

June

26

27

28

29

30

July

1

2

14. Tobacco (*Nicotiana tabacum*) and Nicotine

Nicotiana tabacum, cultivated tobacco, is one of over 70 species in the genus. It is thought to be a hybrid of *Nicotiana sylvestris* (woodland tobacco), *Nicotiana tomentosiformis*[1] and possibly *Nicotiana otophora*. Almost all the genus originates from South America with a single species from Australia and another from Africa.

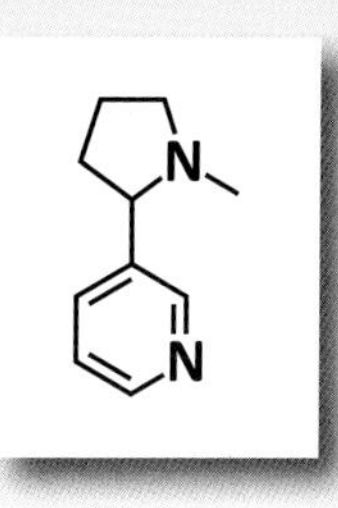

Along with *Nicotiana rustica* (Aztec tobacco), it provided the source of tobacco used in the Mayan civilisation. Taken in large amounts, tobacco has hallucinogenic properties therefore it was often associated with rituals and ceremonies and used by shamans. Extreme doses lower the heart rate and can lead to a catatonic state. For these purposes, tobacco was chewed, licked, eaten, taken as snuff or as an enema but most commonly, smoked. In lower doses, tobacco was used widely for medicinal purposes to disinfect and ward off disease.

Tobacco was introduced to Europe by traders returning from the Americas in the 15th and 16th centuries. At this time, the plant was known

by a variety of names including henbane of Peru, sharing some narcotic properties with henbane. Its virtues were extolled by the French ambassador to Portugal, Jean Nicot de Villemain (1530–1600), who introduced the plant to the French court after witnessing at first hand its medicinal benefits. Tobacco was considered to be a panacea to treat all types of ailments including catarrh, colds and fevers, abscesses and other sores, as a purgative and narcotic and for pain relief. It became known as the ambassador's herb or nicotiane in honour of Nicot. The name tobacco actually applied to the pipe with which it was smoked by the Native Americans (tabaco or tavaco) and was initially used for the plant in error.[2] Nicot also popularised using tobacco as snuff which became known as the Nicotian habit.

The popularity of tobacco meant it became an important crop but some religious leaders considered smoking it to be sinful, in part due to farmland being used to produce this largely recreational substance instead of food crops. James VI of Scotland and I of England was a staunch anti-smoker, writing in 'A Counterblaste to Tobacco' that smoking is 'A custome lothsome to the eye, hatefull to the Nose, harmefull to the braine, dangerous to the Lungs, and in the blacke stinking fume thereof, neerest resembling the horrible Stigian smoke of the pit that is bottomelesse'.[3] He enforced a tax increase of 4000% on tobacco in 1604 but this attempt to curb its use was unsuccessful.

During the outbreak of bubonic plague in London in 1665–6, posies of tobacco were hung outside houses to ward off the disease and tobacco was also smoked for the same purpose. Tobacco was given to residents in a Bolton workhouse in 1882 during an outbreak of smallpox and smokers seemed to be at lower risk of diphtheria and cholera. The effect was probably due to pyridine in the smoke killing germs but the negative effects of tobacco smoking meant its use was not universally supported.[4]

Tobacco smoke contains over 70 carcinogenic compounds (including arsenic, benzene and acreolin) as well as other poisons (including ammonia and hydrogen cyanide). Tar is a collection of hydrocarbon particles which smokers inhale and it forms a sticky brown residue on the teeth, nose and lungs.[5]

While smoking tobacco causes brown staining of the teeth, tobacco paste or creamy snuff which is a combination of tobacco, clove oil, spearmint, menthol and camphor, is sold as a toothpaste in India. A mixture of tobacco powder and molasses is used for the same purpose in Central and Eastern India where it is known as gul. Preparations of roasted and powdered tobacco (mishri or red tooth powder) are also used to clean the teeth.[6] Tobacco is sometimes chewed to relieve toothache.

From the first widely publicised claims of links between smoking and cancer in 1952,[7] there has been a drive to encourage people to stop smoking and it has increasingly been banned in public places. Fidel Castro, closely associated with smoking Havana cigars, gave up smoking in 1985.

Pipe and cigar smoking deliver less nicotine to the body but it is in the form of cigarettes that tobacco is most commonly smoked. Cigarettes are manufactured from cured and dried tobacco leaves and reconstituted tobacco, usually combined with additives. This is rolled into a paper-wrapped cylinder tipped with a filter made of cellulose acetate. They were initially marketed as being good for health, relieving stress and making smokers feel more confident and relaxed. For this reason, British soldiers were supplied with cigarettes during both World Wars.

The compound nicotine makes up 0.6–3% of the dry weight of tobacco. It is a neurotoxin present in the plant to deter predators. It can be used as a pesticide close to harvesting time as it is quickly degraded and is used against piercing or sucking insects such as aphids and whiteflies. *Nicotiana rustica* contains up to 9% nicotine making it an even more effective pesticide.

Nicotine is one of the most addictive substances known to man and can act as a stimulant or a relaxant, depending on concentration. When nicotine-containing smoke is inhaled, there is a release of glucose and adrenaline and a decrease in insulin level which raises the metabolic rate and suppresses appetite. Acetylcholine is released, increasing stimulation, alertness and concentration and enhancing the effect of noradrenaline and dopamine on the brain. Acetylcholine is chemically similar to nicotine therefore its receptors are activated by nicotine. These effects dominate at low blood nicotine concentration. Taking deeper breaths when smoking increases the quantity of nicotine taken in and leads to a depression of nerve transmissions. The action of serotonin is also enhanced, both of which promote relaxation and pain relief.

The action of nicotine to enhance dopamine activity renders it equally or even more addictive than heroin and cocaine.[8] Symptoms of withdrawal include irritability, disrupted sleep, increased appetite and depression. In 2012, the World Anti-Doping Authority placed it on their monitoring list to ascertain if the possible performance-enhancing effects, such as delaying fatigue and elevating mood, should lead to a ban of its use by participants in competitive sports.

Nicotine has some potential medical benefits in treating or delaying the onset of Parkinson's disease and in the prevention and treatment of Alzheimer's disease. It reduces the occurrence of pre-eclampsia and can be used to treat

Woodland tobacco (*Nicotiana sylvestris*)

ulcerative colitis, some forms of epilepsy, attention deficit hyperactivity disorder (ADHD) and obsessive compulsive disorder (OCD).[9] It boosts the growth of blood vessels so could be helpful for diabetics suffering the effects of poor circulation, such as gangrene.[10] However, most of these potential benefits are negated by the detrimental effects if the nicotine is obtained via smoking tobacco.

Cigarette smoking is estimated to lead to 100,000 deaths a year in UK. 8 out of 10 cases of lung cancer can be linked to smoking and it is also related to chronic obstructive pulmonary disease (COPD), heart disease, cancers of the mouth, larynx etc., poor circulation, reduced fertility and more. Smoking during pregnancy increases the risk of miscarriage and can cause defects in the baby and smoking in the presence of others can affect their health too.[11]

References

1. J.C. Gray, S.D. Kung, S.G. Wildman, S.J. Sheen, (1974); *Nature*, 252, 226–7, doi: 10.1038/252226a0
2. A. Charlton, (2004); *Journal of the Royal Society of Medicine*, 97(6), 292–6
3. Digitised translation, (2002); http://www.laits.utexas.edu/poltheory/james/blaste/ [Accessed 16.11.2013]
4. Anonymous, (1889); *British Medical Journal*, 2, 247–64, (p 253), doi: 10.1136/bmj.2.1492.247
5. http://www.cancerresearchuk.org/cancer-info/healthyliving/smokingandtobacco/whatsinacigarette/ [Accessed 16.11.2013]
6. Smokeless Tobacco Fact Sheets, (2002); http://cancercontrol.cancer.gov/brp/TCRB/stfact_sheet_combined10-23-02.pdf [Accessed 16.11.2013]
7. R. Norr, (December 1952); *Reader's Digest*, 7–8
8. S. Blakeslee, (1987); New York Times, http://www.nytimes.com/1987/03/29/magazine/nicotine-harder-to-kickthan-heroin.html [Accessed 16.11.2013]
9. W.E. Leary, (1997); New York Times, http://www.nytimes.com/1997/01/14/science/researchers-investigate-horrors-nicotine-s-potential-benefits.html?pagewanted=all&src=pm [Accessed 17.11.2013]
10. J. Jacobi, J.J. Jang, U. Sundram, H. Dayoub, L.F. Fajardo, J.P. Cooke, (2002); *American Journal of Pathology*, 161(1), 97–104, doi: 10.1016/S0002-9440(10)64161-2
11. http://www.patient.co.uk/health/smoking-the-facts [Accessed 17.11.2013]

July

3

4

5

6

7

8

9

July

10

11

12

13

14

15

16

15. Foxglove (*Digitalis* spp.) and Digoxin

Digitalis is a genus of about 20 species commonly known as foxgloves, the most abundant of which is *Digitalis purpurea*. Flowers can fit over a human finger hence the name digitalis. The name is also commonly used to describe a group of chemical compounds known as cardiac glycosides extracted from the plants which can be used to treat heart conditions. The most common of these are digoxin and digitoxin. They are known as glycosides as they contain sugar groups within the molecule.

Digitalis has been used medicinally since the 13th century to help heal wounds, clear phlegm and to treat cases of 'the king's-evil' (scrofula). It was also used as a diuretic and to treat epilepsy. However, the herbalist John Gerard, although noting these useful properties, dismissed it as having no place amongst medicines, perhaps due to the whole plant being toxic.

The introduction of digitalis into Western conventional medicine can be attributed to William Withering (1741–1799), a Shropshire physician, botanist, chemist and geologist. In 1776, he met an old woman who used a preparation of over 20 herbs to successfully treat dropsy. Dropsy, now known as oedema, is the swelling of tissues due to the build-up of fluid and is often related to a weak heart and inefficient kidney function. Over the next 9 years, Withering identified the active herb in the mixture as the foxglove and documented 156 cases where he used it, determining the dosage and best practise for collecting, extracting and administrating the active part.[1]

In 1779, Withering started work at Birmingham General Hospital where Erasmus Darwin, grandfather of Charles, worked. At least one of Withering's patients was referred to him by Darwin and in 1780, Darwin reported on the therapeutic use of foxglove, claiming priority for the development of a new treatment for pulmonary consumption for himself and his deceased son, also Charles. Although both Withering and Darwin were members of the prestigious Lunar Society, their relationship deteriorated with Darwin accusing Withering of professional malpractice.[2] He was also in conflict with Withering over their botanical texts, Withering's Botanical Arrangement of all the Vegetables naturally growing in Great Britain published in 1776 and Darwin's translation of Linnaeus' Genera Plantarum and Mantissa Plantarum, 'The Lichfield Linnaeus', published in 1787. One disagreement was over the use of sexualised language in the translation of Linnaeus – while Withering believed ladies should be protected from it, Darwin believed

Linnaeus should be translated as literally as possible. Although Darwin won this particular argument, it is Withering who is credited with the discovery of digitalis.

The action of digitalis is to slow the heart rate but also to strengthen it. These compounds affect the 'sodium pump' in the heart, inhibiting sodium-potassium ATP-ase, resulting in more sodium ions in the cells. The knock-on effect is that the calcium ion concentration is also increased and the heart becomes more contractive. It also acts on the vagus nerve and the heart rate slows.

However, it has a low therapeutic index – this is the ratio of the dose which has a lethal effect to the dose which has a positive therapeutic benefit. As plants produce differing amounts of active compounds depending on location, when they are harvested etc., using specific compounds extracted from the plant and purified will give a more reliable dosage and increase the likelihood of a positive outcome. Digitalis is no longer used in herbal medicine.

The compound digoxin is usually obtained for medicinal use from the woolly (or Grecian) foxglove (*Digitalis lanata*). It can be used to treat atrial fibrillation, atrial flutter and heart failure which cannot be treated by other methods. In most cases, alternative treatments now exist so digoxin is not widely used. The action of the compound digitoxin is similar but the effects are longer lasting. Unlike digoxin which is eliminated from the body via the kidneys, digitoxin is eliminated via the liver therefore it is useful for patients with poor kidney function.

Other side effects of the use of digoxin include loss of appetite, nausea, vomiting and visual disturbances. The artist Vincent van Gogh (1853–90) is believed to have taken digitalis for epilepsy and this may have caused him to see with a yellow-green tinge, with coronas surrounding yellow objects such as the stars. His view of the world is demonstrated in the 1889 painting 'Starry Night'. The idea is supported by the portrait of his physician Paul-Ferdinand Gachet in 1890, who is shown holding *Digitalis purpurea*.[4] However, van Gogh had a range of medical issues for which he may have been undergoing treatment and which may have caused or contributed towards the problem.

Another compound obtained exclusively from foxgloves is the steroid digoxigenin. This is a hapten, a small molecule which causes an

immune response only when attached to a larger molecule like a protein. It can therefore be used as a marker to allow study of biological processes.[5]

The whole of the foxglove plant is poisonous leading to it being given a range of common names like dead men's bells, bloody fingers and witches' thimbles. The spots on the blossoms were said to be made by elves' fingers and warn of the poisonous juices. The plant or its chemical components have been used as a weapon of murder both in real life and in fiction. Charles Cullen, an American nurse, was convicted of killing 40 patients using either insulin or digoxin over a 16 year period. He was sentenced to multiple life sentences in 2006 and experts believe he may have killed over 300 patients.[6] In 2010, Lisa Leigh Allen of Colorado, USA, was convicted of trying to poison her husband by giving him a salad containing foxglove leaves. She was sentenced to 4½ years in jail.[7] In fiction, digitalis was used as a murder weapon six times by Agatha Christie in her novels as well as by Dorothy Sayers and Mary Webb. Digitalis has also been used for fraudulent purposes. In the 1930s, a group of doctors and lawyers tried to defraud insurance companies by simulating heart disease in patients using digitalis.[8] There have been deaths caused by mislabelling of pharmaceutical preparations in both Belgium and Holland.[8]

Growing foxgloves near other plants protects these from disease and in the case of apples, potatoes and tomatoes, improves their storage behaviour. As a cut flower, foxgloves prolong the life of other flowers in the vase and, alternatively, a tea can be made from the stems and flowers and added to the water for the same effect.

References

1. W. Withering, (1785); *An Account of the Foxglove and Some of its Medical Uses with Practical Remarks on Dropsy and other Diseases*, CreateSpace Independent Publishing Platform (May 2013), ISBN: 1484938240
2. G.C. Cook, (1999); *Journal of Medical Biography*, 7(2), 86–92
3. D.M. Krikler, (1985); *Journal of the American College of Cardiology*, 5(5), Supplement 1, 3A–9A, doi: 10.1016/S0735-1097(85)80457-5
4. P. Wolf, (2001); *Western Journal of Medicine*, 175(5), 348
5. S.M. Hart, C. Basu, (2009); *Journal of Biomolecular Techniques*, 20(2), 96–100
6. C. Graeber, (2007); New York Magazine; http://nymag.com/news/features/30331/ [Accessed 19.11.2013]
7. http://www.fox21news.com/news/story.aspx?id=457169#.UotNR8SpWSo [Accessed 19.11.2013]
8. H.B. Burchell, (1983); *Journal of the American College of Cardiology*, 1(2-1), 506–16 doi: 10.1016/S0735-1097(83)80080-1

July

17

18

19

20

21

22

23

July

24

25

26

27

28

29

30

16. Common Milkweed (*Asclepias syriaca*) and β-Amyrin

The milkweeds are a genus of over 140 species including the common milkweed, *Asclepias syriaca*. They share an unusual method of pollination – rather than existing as individual grains, pollen is held together by sticky threads to form sacs known as pollinia. As an insect feeds on nectar, its legs slip into one of the slits in each flower formed by adjacent anthers. At this point, a pair of pollen sacs detach from the plant and attach to the insect's leg. When the insect visits the next flower, the process is reversed and the sacs are released. This method of pollination does not always have a good outcome for the insects. Honeybees have been found starved to death with multiple sacs which have become entangled, attached.[1] Milkweed nectar has a high dextrose content and was traditionally used as a sweetener by Native Americans. In the 18th century, French Canadians made a type of brown sugar from the flower heads.

Most milkweeds contain a milky juice from which they get their common name. This clots on exposure to air similar to the case for spurges. The primary role for this is defence with the sticky latex gumming up the mouths of insects and also containing emetic compounds. Other defences are hairs on the leaves and cardenolides such as syrioside in the tissues. Cardenolides and related cardiac glycosides act to slow down the heart rate and strengthen the beats similar to digitalis. Milkweeds also protect

neighbouring plants from some pests including click beetle larvae.

The sap contains a number of chemical compounds, including α- and β-amyrin. A mixture of these closely related tri-terpenes has been shown to possess anti-hyperglycaemic and hypolipidemic effects and could potentially be developed for use in the treatment of diabetes and atherosclerosis.[2] β-Amyrin also has anti-bacterial, anti-fungal, anti-inflammatory and anti-cancer properties.[3]

Some potential uses of milkweed stems and leaves have been investigated. Their viability differed with species due to differing chemical composition. *Asclepias speciosa* contained mainly α- and β-amyrin while *Asclepias curassavica* contained mainly cardiac glycosides. The former has a higher calorific value than the latter therefore has greater potential as a biofuel but both have potential as chemical feedstocks.[4]

Milkweeds are the only food source for monarch butterfly larvae. These, along with other insects feeding on the plants, have evolved strategies to circumvent the plants' defences. Some caterpillars are immune to the cardenolides. Others shave the leaves and cut the plant's veins in a circle, eating the centre where the latex doesn't flow, or alternatively, cut the mid-vein, stopping latex flow to the rest of the leaf.

Insects eating milkweeds, themselves become protected from predation by the milkweed toxins. Birds eating monarch larvae feel nauseous and vomit and the effect is retained when the larvae become butterflies. Perhaps as a result of some predators being immune to milkweed's defences, chemical defence is becoming a less important strategy for protection, being replaced by faster regrowth. Milkweeds are evolving to grow more quickly than the caterpillars can eat them which may be more energy efficient.[5]

The milky sap from most species of milkweed contains caoutchouc, natural rubber. Thomas Edison experimented on making rubber from this sap. Although he was successful in producing low-grade material, this was not pursued again until World War II when shortage of rubber re-ignited interest in both USA and Germany. However, the discovery of a method for producing man-made rubber meant this promise was never fulfilled. The sap has been used topically to treat warts, ringworm, scrofula and bee stings. It can cause mild dermatitis therefore gloves need to be worn when handling the plants.

Milkweed seeds are produced in follicles covered in hairs known as silk or floss. This floss was traditionally used as the background for mounted butterflies. The follicles are hollow and coated in wax so have good insulation properties. These are used commercially as a filling for hypoallergenic pillows and quilts, and also

in jackets, sometimes mixed with down. The follicles are waterproof and buoyant – about 6 times more buoyant than cork. Kapok, a fibre from the kapok tree (*Ceiba pentandra*), is very buoyant and supplies from Java were used to fill life-jackets. When the island was occupied by the Japanese during World War II, children in USA were paid to collect milkweed seed pods and the follicles used as an alternative filling material.[6]

The bast fibre, collected from the inner bark coating the stem, was favourably evaluated by the US Department of Agriculture in the 1890s and again in the 1940s, as an alternative to hemp and flax. It was used by Native Americans to make rope or coarse fabric and it can also be used to make paper. However, these potential uses of milkweed have never been commercially developed.

Native Americans made dyes from milkweeds and ate several parts of the plant – carefully prepared to remove the toxins. Some foragers eat milkweed to this day and report that the tender shoots taste like asparagus. However, milkweeds are toxic and there have been deaths of livestock and pets caused by ingesting plant material. Milkweed toxins have been used as poisons for arrow tips in South America.

The oil extracted from the seeds can be reacted to give cinnamic acid derivatives which offer a high degree of UV-protection at low concentration (1–5%). Added to the oil's other components including anti-oxidants such as tocopherols, this offers the opportunity of developing milkweed-derived sun-screens and other skin and hair care products.[7] The oil is rich in omega-7 fatty acids which add elasticity to the skin, and also in vitamin E. Oil is sold in the form of a balm applied as an anti-inflammatory and suggested as a possible food supplement. However, the presence of toxins in the plant is problematic to gaining approval for this latter purpose.[6]

The genus *Asclepias* was named after the Greek god of healing, Asclepius. The rod of Asclepius, a snake-entwined staff, is the symbol of medicine. In common with other members of the genus, *Asclepias syriaca* is native to North America. It was included in French botanist and physician Jacques-Philippe Cornut's 1635 book of North American plants. Although he had confused it with a species native to Syria, the name syriaca was retained by Carl Linnaeus. Plants from this genus have a wide range of uses in traditional medicine.

Asclepias tuberosa is known as pleurisy root or butterfly weed. It produces copious quantities of orange flowers and nectar which attract butterflies, hummingbirds and other insects. In herbal medicine, the dried root is used against respiratory infections where it acts as an anti-inflammatory and expectorant. In addition, its anti-spasmodic and diaphoretic properties (promoting toxin removal by sweating) mean it is

Butterfly weed (*Asclepias tuberosa*)

useful in the treatment of pleurisy, pneumonia and flu.[8] A study of pleurisy root showed extracts to possess oestrogenic and uterine-stimulating properties.[9] Unlike other members of the milkweed genus, pleurisy root contains no (or almost no) milky sap.

A decoction from common milkweed has been used as an anti-fertility treatment, to promote milk-flow after childbirth and to prevent haemorrhage. It can be used as a laxative, stomach medicine and for dropsy.[10]

Other members of the genus used in traditional medicine include *Asclepias verticillata* (whorled milkweed) which was used in the southern USA to treat snake bites and the bites of venomous insects. *Asclepias curassavica* (scarlet milkweed) was used as an emetic and the juice as an anthelmintic (to destroy worms) for children and to repel fleas.[11] A recent study has shown extracts of this species to have anti-bacterial and anti-fungal activity.[12]

References

1. J.F. James, (1887); *The American Naturalist*, 21(7), 605–15, online via JSTOR at http://archive.org/stream/jstor-2451222/2451222#page/n1/mode/2up [Accessed 20.11.2013]
2. F.A. Santos, J.T. Frota, B.R. Arruda, T.S. de Melo, A.A. de Carvalho, A. da Silva, G.A. de Castro Brito, M.H. Chaves, V.S. Rao, (2012); *Lipids in Health and Disease*, 11, 98, doi: 10.1186/1476-511X-11-98
3. L.H. Vázquez, J. Palazon, A. Navarro-Ocaña, (2012); *A Review of Sources and Biological Activities in Phytochemicals - A Global Perspective of Their Role in Nutrition and Health*; Chapter 23: The Pentacyclic Triterpenes α- and β-amyrins, Ed. V. Rao, InTech, ISBN 978-953-51-0296-0, doi: 10.5772/27253
4. J. Van Emon, J.N. Seiber, (1985); *Economic Botany*, 39(1), 47–55, doi: 10.1007/BF02861174
5. A.A. Agrawal, M. Fishbein, (2008); *Proceedings of the National Academy of Sciences*, 105(29), 10057–60, doi: 10.1073/pnas.0802368105
6. P. Clark, (2012); The Washington Post, http://www.washingtonpost.com/wp-srv/special/metro/urban-jungle/pages/120925.html [Accessed 20.11.2013]
7. J. Suszkiw, (2009); *Agricultural Research*, February 2009, p 9, http://www.ars.usda.gov/is/AR/archive/feb09/milkweed0209.pdf [Accessed 20.11.2013]
8. D. Hoffmann, (1996); *Complete Illustrated Guide to the Holistic Herbal*, Element Books, ISBN 0-00-713301-4, p 66
9. C.H. Costello, C.L. Butler, (1950); *Journal of the American Pharmaceutical Association*, 39(4), 233–7 doi: 10.1002/jps.3030390414
10. USDA Plants Database, http://plants.usda.gov/plantguide/pdf/cs_assy.pdf [Accessed 21.11 2013]
11. Mrs M. Grieve, (1973 edition); *A Modern Herbal*, Merchant Book Company Ltd., ISBN 1904779018 pp 64–5
12. C. Hemavani, B. Thippeswamy, (2012); *Recent Research in Science and Technology*, 4(1), 40–3, http://recent-science.com/index.php/rrst/article/view/11015/5568 [Accessed 21.11.2013]

July

31

August

1

2

3

4

5

6

August

7

8

9

10

11

12

13

17. Feverfew (*Tanacetum parthenium*) and Melatonin

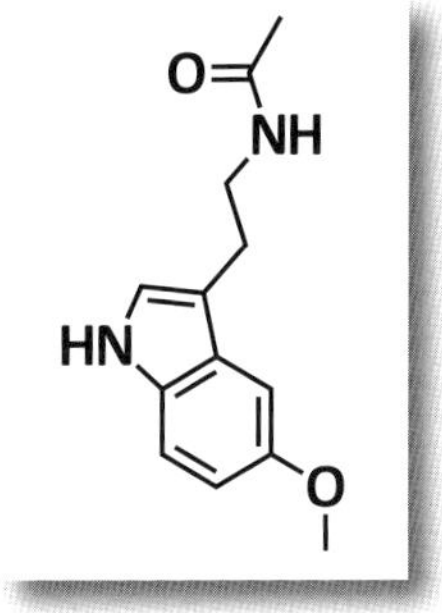

Feverfew (*Tanacetum parthenium*) has a long history of medicinal use as shown by both its Latin and common names. Legend has it the plant was used to help save the life of someone who had fallen from the Parthenon in Athens, earning it a species name in recognition. The medicinal virtues were noted by the Greek physician Dioscorides in the first century AD. He recommended its use for many ailments including inflammation, arthritis, phlegm, swelling, headaches and fever. The name feverfew is a corruption of febrifuge, fever-reducer. Feverfew was also long associated with the treatment of 'women's complaints' and an alternative explanation for the name parthenium is that it derives from the Greek word *panthenos*, meaning virgin. Feverfew should not be taken during pregnancy as it may cause contractions to begin.

Other common names include wild chamomile, midsummer daisy and bachelor's buttons (along with some other species). Feverfew is found growing in hedgerows but is also cultivated as a garden plant. It is a member of the daisy family (Compositae) which also includes the chamomiles and it produces daisy-like flowers. The yellow flower heads are nearly flat discs unlike the conical flower heads of chamomile. In feverfew, white rays radiate from this flower head, while in the related species tansy (*Taracetum vulgare*) these are absent. Tansy was traditionally a strewing herb and insecticide. Bodies were sometimes packed with tansy to preserve them until burial. Its common name relates to the latter practice, coming from the Greek word *athanasia* meaning immortality.

Feverfew was used in traditional medicine as an anti-histamine to treat allergies and relieve asthma, taken hot to reduce fevers and as a decongestant as well as to treat headaches and inflammation. There has been a considerable amount of work undertaken to identify and understand the origin of the medicinal properties of this plant.

Feverfew contains the flavanol tanetin which is thought to contribute to its anti-inflammatory activity.[1] Another compound of biological significance is parthenolide, a sesquiterpene lactone. Parthenolide is a skin-sensitiser. One traditional method of administering feverfew was to eat the bitter leaves in a sandwich, making them more palatable. However, the leaves cause irritation and ulcers in sensitive people. A study looking at the possible use of parthenolide-depleted feverfew showed this has anti-inflammatory activity without the negative effects of sensitisation.[2]

Feverfew inhibits platelet aggregation in the bloodstream preventing the blockage of small capillary blood vessels. This behaviour has been attributed to parthenolide and may contribute to feverfew's success in treating headaches.[3] However, it gives feverfew blood-thinning properties so it should not be taken with warfarin or other blood-thinning agents.

Feverfew's anti-inflammatory and anti-cancer properties have also been attributed to parthenolide. Of particular importance is the finding that parthenolide does not act on healthy cells, making it a prominent candidate for use in the treatment of cancer and inflammation-related disorders.[4] Parthenolide blocks the release of serotonin and prostaglandin and acts as a vasodilator, decreasing menstrual and headache pain. Serotonin is related to another compound found in abundance in feverfew called melatonin.

Tansy (*Tanacetum vulgare*)

Melatonin is a hormone found widely in the plant and animal kingdoms. In plants, one role is to regulate response to the length of day and night (photoperiod). Responses include flowers opening, stems and roots growing and leaves falling. It can also slow root growth while promoting growth above ground. Melatonin also governs the response to harsh conditions.[5]

In animals including humans, melatonin is secreted by the pineal gland in the brain during darkness, leading to its nickname, the 'hormone of darkness'. The duration of the secretion depends on the length of the day and night so varies seasonally. In some animals, it gives a biological signal for seasonal functions such as reproduction (reducing libido during winter), coat growth and shedding, development of camouflage etc. It promotes sleep by lowering the core body temperature and

causing drowsiness.[6] This has led to its use to treat circadian sleep disorders and jet-lag and it is known as a chronobiotic. The amount of melatonin produced decreases with age and there is a significant change in melatonin production around puberty. This may partially account for teenagers' sleep patterns – melatonin is produced later in the day, leading to later sleeping and waking times.[7]

Pineal production of melatonin is suppressed by light, particularly in the blue part of the spectrum. Blocking blue light by wearing amber lenses has been shown to improve the quality of sleep and major changes to sleep times can be facilitated by omitting blue light for several hours before sleep is desired. As mood is affected by the quality of sleep, melatonin also has a mood-improving effect.[8]

Melatonin is a very effective anti-oxidant, unusual in that it is amphiphilic i.e. it can be mixed with both fats and non-fats and also for the range of metabolites it produces which also have anti-oxidant properties. This may have been melatonin's primary function in nature, its other effects developing later in the evolutionary process.[9] Melatonin also improves the effectiveness of other anti-oxidants. Its free-radical scavenging properties mean it gives protection against radiation.

Melatonin has proven effective in the treatment of migraines and some other types of headache. This may be partly due to its anti-oxidant and anti-inflammatory effects.[10] It has been linked to memory and learning and is of interest as a potential treatment for the effects of Alzheimer's disease.[11] Other potential medicinal uses are in the treatment of immunodeficiency states and cancer, as an anti-viral agent,[12] for the treatment of irritable bowel syndrome (IBS), obesity and tinnitus, to reduce the occurrence of Type 2 diabetes and to increase the mobility of gallstones.

Melatonin is involved in the mechanism by which some reptiles and amphibians e.g. chameleons, change colour. In low concentrations it activates the Mel_{1C} receptors, causing melanin-containing pigment granules (melanosomes) in melanophores to move to the centre of the cells and the skin colour to lighten.[13] Chromatophores are pigment containing cells responsible for skin colour. Melanophores are black chromatophores which govern how much light is reflected.

Feverfew tea was a traditional disinfectant with insect repelling properties. When grown in the garden, feverfew protects nearby plants from insects but as it repels bees, it should not be planted near any species which require bees for pollination.[14] Common to other members of this genus, it contains pyrethrins, natural

insecticides. One of the most abundant sources of pyrethrins is pyrethrum (*Tanacetum cinerariifolium, Chrysanthemum cinerariifolium* or Dalmatian chrysanthemum). Pyrethrum also refers to the insecticide made from the leaves of the plant. It is considered one of the safest insecticides for use on food crops as it is biodegradable and rapidly decomposes in air and sunlight. It is also used in flea powder for pets.

References

1. C.A. Williams, J.R.S. Hoult, J.B. Harborne, J. Greenham, J. Eagles, (1995); *Phytochemistry*, 38(1), 267–70, doi: 10.1016/0031-9422(94)00609-W
2. R. Sur, K. Martin, F. Liebel, P. Lyte, S. Shapiro, M. Southall, (2009); *Inflammopharmacology*, 17(1), 42–9, doi: 10.1007/s10787-008-8040-9
3. W.A. Groenewegen, S. Heptinstall, (1990); *Journal of Pharmacy and Pharmacology*, 42(8), 553–7, doi: 10.1111/j.2042-7158.1990.tb07057.x
4. V.B. Mathema, Y.-S. Koh, B.C. Thakuri, M. Sillanpää, (2012); *Inflammation*, 35(2), 560–4, doi: 10.1007/s10753-011-9346-0
5. M.B. Arnao, J. Hernández-Ruiz, (2006); *Plant Signalling and Behaviour*, 1(3), 89–95, doi: 10.4161/psb.1.3.2640
6. J. Arendt, D.J. Skene, (2005); *Sleep Medicine Reviews*, 9(1), 25–39, doi: 10.1016/j.smrv.2004.05.002
7. http://kidshealth.org/teen/your_body/take_care/how_much_sleep.html# [Accessed 24.11.2013]
8. K. Burhhart, J.R. Phelps, (2009); *Chronobiology International*, 26(8), 1602–12, doi: 10.3109/07420520903523719
9. D.-X. Tan, L.C. Manchester, M.P. Terron, L.J. Flores, R.J. Reiter, (2007); *Journal of Pineal Research*, 42(1), 28–42, doi: 10.1111/j.1600-079X.2006.00407.x
10. M.F.P. Peres, M.R. Masruha, E. Zukerman, C.A. Moreira-Filho, E.A. Cavalheiro, (2006); *Expert Opinion on Investigational Drugs*, 15(4), 367–75, doi: 10.1517/13543784.15.4.367
11. X.-C. Wang, J. Zhang, X. Yu, L. Han, Z.-T. Zhou, Y. Zhang, J.-Z. Wang, (2005); *Acta Physiologica Sinica*, 57(1), 7–12
12. G.J.M. Maestroni, (1999); *Tryptophan, Serotonin, and Melatonin, Advances in Experimental Medicine and Biology*, 467, 217–26, doi: 10.1007/978-1-4615-4709-9_28
13. D. Sugden, K. Davidson, K.A. Hough, M.-T. Teh, (2004); *Pigment Cell Research*, 17(5), 454–60, doi: 10.1111/j.1600-0749.2004.00185.x
14. T. Breverton, (2011); *Breverton's Complete Herbal*, Quercus Publishing Ltd, ISBN 978-0-85738-336-5, p 147

August

14

15

16

17

18

19

20

August

21

22

23

24

25

26

27

18. Sweet Wormwood (*Artemisia annua*) and Artemisinin

The genus *Artemisia* contains between 200 and 400 species including *Artemisia dracunculus* (tarragon) which is used in cookery. Common wormwood (*Artemisia absinthium*) is used to flavour the drink absinthe while Roman wormwood (*Artemisia pontica*) is used to flavour the liqueur vermouth. The name vermouth derives from the German word for wormwood, *Wermut*. Most species have strong aromas and a bitter taste to discourage herbivores. 'As bitter as wormwood' was a common saying and wormwood is a byword in the Bible for bitterness and sorrow.

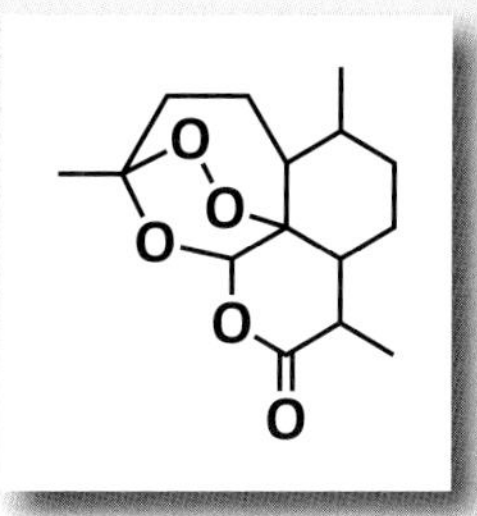

Artemesia annua (sweet wormwood, annual wormwood or sweet Annie) has fern-like leaves, an abundance of small yellow flowers and a scent like camphor. It is native to Asia and has been used in traditional Chinese medicine (TCM) to treat intermittent fevers (malaria) and skin diseases since 200 BC. Both *Artemisia annua* and *Artemisia apiacea hance* are known in TCM as qinghao but the former is more correctly known as huanghuahao.

The medicinally important constituent of qinghao is artemisinin or qinghaosu – 'basic element of qinghao' in Chinese. This was identified during plant screening studies to find a treatment for the malaria-ridden North Vietnamese army, China's allies, as part of Project 523.[1] This project commenced in 1967 during the Cultural Revolution and involved around 500 scientists at 60 institutions across China. Scientists worked in secret, often underground and at night to avoid the attention of the Red Guards.

Malaria is caused by the mosquito-borne parasite *Plasmodium* and is widespread in areas with

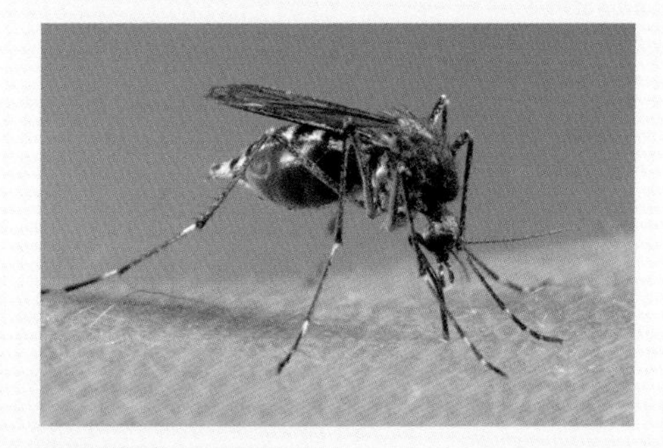

Absinthe

a summer temperature of 27–28°C. In 2010, there were an estimated 219,000,000 cases and 660,000 deaths worldwide. Among the most common and deadly of the *Plasmodium* parasites are *Plasmodium falciparum* and *Plasmodium vivex*.[2]

Although qinghaosu was discovered in 1972, it was kept secret until 1979 when Chinese reform allowed its reporting outside the country.[3] In the 1980s, the World Health Organisation (WHO), working with Chinese scientists, tried to bring artemisinin into use worldwide but disputes led to the end of this collaboration in 1985. It was not until 1999, after the intervention of the pharmaceutical company Novartis, that a malaria treatment based on artemisinin reached the world market. The discovery of artemisinin has since been described as 'the greatest medical discovery in the second half of the 20th century', by Professor Richard Haynes and has saved millions of lives.[4]

During the Vietnam War, it was not only the North Vietnamese that struggled with drug-resistant malaria among troops. America's response was to instigate research at the Walter Reed Army Institute of Research into new drugs for malaria. This led to the discovery of mefloquine, a synthetic analogue of quinine. Quinine, obtained from the bark of the tropical *Cinchona* trees, has been used to treat malaria in the West since the early 17th century. Attempts to chemically synthesise quinine resulted serendipitously in the discovery of the first synthetic organic dye, mauveine, in 1856. Malaria parasites have increasingly developed resistance to quinine and this, along with the possible side-effects of taking it, led to the recommendation by WHO in 2006 that quinine should no longer be used as a first-line treatment for malaria except in special circumstances.

Artemisinin is a sesquiterpene lactone with an endo-peroxide (O-O) bridge. Bridging peroxide groups are very rare in naturally-occurring compounds and this may account for the scepticism shown by Western scientists when it was first reported. Another notable example is the bicyclic terpene ascaridole, obtained from epazote or wormseed (*Dysphania ambrosioides*), which is an anthelmintic (expels parasitic worms). Importantly, artemisinin belongs to a totally new structural type among anti-malarial drugs, therefore it was thought likely that malaria parasites would not quickly become resistant.

The peroxide group in artemisinin is believed to be crucial to its activity. It renders the molecule very reactive to iron, present in haem, a component of the red pigment in blood cells where malarial parasites mainly reside and in other proteins such as myoglobin. There has been much debate about the exact mode of action but a recent paper reconciles two different theories into a single process. It suggests artemisinin interacts with iron, breaking the O-O bond

forming an iron adduct. This interferes with the parasite's calcium pump (PfATP6), blocking its action and ultimately killing it.[5]

However, use of artemisinin is not without problems. As in the case of quinine, malaria parasites have begun to develop immunity. In 2006, WHO asked pharmaceutical companies to stop marketing artemisinin as a monotherapy for malaria but it is still used this way in some countries. Another issue is the sale of fake or sub-standard drugs. These often contain less than the required dose of the active ingredient so do not cure the disease, instead promoting the development of drug-resistance. In some areas, it is estimated that over 50% of anti-malaria drugs are counterfeit or sub-standard.[6]

Although artemisinin is effective in destroying malaria parasites, it has a short lifetime in the body and full clearance of all parasites is not obtained. As a result, it is often combined with other anti-malaria drugs with longer lifetimes, such as mefloquine. Artemisinin combination therapy (ACT) is the recommended first-line treatment against *Plasmodium falciparum* malaria. Artemisinin is also effective against *Plasmodium vivex* malaria and against cerebral malaria, the most serious form.

Artemisinin is poorly soluble in water which reduces its bioavailability. Many derivatives have, therefore, been synthesised and tested for their activity against malaria. The simplest of these is dihydroartemisinin (artenimol), the result of adding two hydrogen atoms to artemisinin. It is more soluble and more effective than artemisinin against malaria. Other active derivatives include artesunate, artemether and artelinic acid. Like artemisinin, these are used in combination with other anti-malarial agents.

The poor solubility of the qinghaosu was appreciated by scholars of traditional Chinese medicine. The physician Ge Hong in the 4th century CE, recommended soaking the fresh plant in cold water, wringing it out and ingesting the juice rather than the more usual method of making a tea from the plant.[7]

Apart from treating malaria, artemisinin has other medicinal uses. Cancer cells have a higher level of free iron than normal cells and, similarly to the behaviour during malaria treatment, artemisinin interacts readily with the iron. This produces free radicals which kill the cancer cells. Artemisinin derivatives with improved bioavailabilty and longer half-lives are being tested for their anti-cancer properties.[8]

Tarragon (*Artemisia dracunculus*)

As well as their activity against malarial parasites, artemisinin and its derivatives are also active against schistosome parasites which are the second most common cause of parasitic infection.[9] A topical preparation of artemisinin or its derivatives was patented for the treatment of psoriasis, UV-light-induced skin conditions, tumours, haemorrhoids and some other skin diseases but the patent has now lapsed.[10]

Sweet wormwood is grown commercially in China, Vietnam and East Africa, taking about 8 months to reach harvestable size from seedlings. Researchers working at the National Institute of Agricultural Botany (NIAB) in Cambridge have developed higher yielding *Artemisia annua* which produces three times the industry standard of artemisinin. They are also trying to improve the extraction of artemisinin from the plant material.[11] The quantity of artemisinin in the plant varies due to environmental factors and harvest fluctuates, therefore, to ensure supply, attempts have been made to develop synthetic or semi-synthetic preparations.

Artemisinic acid, a precursor in the synthesis of artemisinin, can be successfully prepared from glucose using genetically modified yeast. This is then converted to artemisinin in four steps, including a photochemical step to generate the peroxide bridge. Industrial scale production by this method began in 2013. A waste product from sweet wormwood processing, dihydroartemisinic acid, can also be converted to artemisinin in good yields, in a continuous-flow process, also involving a photochemical step.[12] These processes, along with other developments, are aimed at providing a reliable and stable supply of the drug.

References

1. Y. Tu, (2011); *Nature Medicine*, 17, 1217–20, doi: 10.1038/nm.2471
2. World Health Organisation, Factsheet number 94, Reviewed March 2013, http://www.who.int/mediacentre/factsheets/fs094/en/ [Accessed 26.11.2013]
3. Qinghaosu Antimalaria Coordinating Research Group, (1979); *Chinese Medical Journal*, 92(12), 811–16
4. J. Zhang, (2013); *A Detailed Chronological Record of Project 523 and the Discovery and Development of Qinghaosu (Artemisinin)*, Translated by M. and K. Arnold, Strategic Book Publishing and Rights Co., ISBN 978-1-62212-164-9, p xxvii
5. A. Shandilya, S. Chacko, B. Jayaram, I. Ghosh, (2013); *Scientific Reports*, 3, Article number: 2513, doi: 10.1038/srep02513
6. http://www.who.int/mediacentre/news/releases/2006/pr02/en/ [Accessed 26.11.2013]
7. E. Hsu, (2006); *Transactions of the Royal Society of Tropical Medicine and Hygiene*, 100(6), 505–8, doi: 10.1016/j.trstmh.2005.09.020
8. H.C. Lai, N.P. Singh, T. Sasaki, (2013); *Investigational New Drugs*, 31(1), 230–46, doi: 10.1007/s10637-012-9873-z
9. S.-H. Xiao, (2005); *Acta Tropica*, 96(2–3), 153–67, doi: 10.1016/j.actatropica.2005.07.010
10. C.R. Thorfeldt, (1990); *US Patent*, US 4978676 A
11. http://www.niab.com/news_and_events/article/109 [Accessed 26.11.2013]
12. M. Peplow, (2013), *Chemistry World*, http://www.rsc.org/chemistryworld/2013/04/sanofi-launches-malaria-drug-production

August

28

29

30

31

September

1

2

3

September

4

5

6

7

8

9

10

19. Globe Artichokes (*Cynara scolymus*) and Cynaratriol

Globe artichokes (*Cynara scolymus*, sometimes *Cynara cardunculus* var. *scolymus*) are a natural variety of the cardoon (*Cynara cardunculus*), members of the thistle family. The flower buds and stems are edible and are harvested before the flowers bloom. Artichokes were grown by the Greeks and Romans and are one of the oldest cultivated vegetables.

The genus name, *Cynara*, is sometimes used as a girl's name and was a character in the poem Non Sum Qualis eram Bonae Sub Regno Cynarae by English poet Ernest Dowson (1867–1900). Margaret Mitchell (1900–1949) used a phrase from the line "I have forgot much, Cynara! gone with the wind"[1] as the title of her novel Gone with the Wind. Cynara was also a young maiden with whom Zeus fell in love. When she wouldn't leave her earthly home to become a goddess, Zeus changed her into an artichoke, with her tender heart surrounded forever by a thorny crown of leaves.[2]

Artichokes became fashionable during the Renaissance when Catherine de Medici (1519–1589) married the French King Henri II and reintroduced the forgotten vegetable. Catherine loved artichokes, believing them

to be an aphrodisiac. The herbalist Culpeper reported they 'provoke lust.'[3] Traditionally, they were also considered to be a contraceptive. Not only did Catherine eat them herself but she encouraged her entourage to do so too, particularly *l'escadron volant* (the flying squadron), her beautiful ladies in waiting who travelled with Catherine wherever she went. Their role was to socialise with the nobles, making them easier for Catherine to deal with and possibly to extract information from them via seduction.

Artichokes spread to England and were popular in the court of King Charles I and his French wife Henrietta Maria who had a large garden of artichokes at her Wimbledon home.[3] The centre of the US artichoke industry is Castroville in California and in 1948, Marilyn Monroe became honorary 'Artichoke Queen' appearing in a sash 'California Artichoke Queen' holding an artichoke. After this, the popularity of artichokes soared in the US.[2]

However, not all connections with the vegetable were positive. New York gangster and onetime deputy of the Morello crime family, Ciro 'The Artichoke King' Terranova (1888–1938) bought the vegetables in California and sold them in New York for 30–40% profit. In 1935, sale or possession of baby artichokes, the kind favoured by Italian immigrants, was banned in New York by Mayor Fiorello LaGuardia, calling the situation 'a serious and threatening emergency'. The aim was to break the Morello family monopoly which brought them around a million dollars a year and give vendors the opportunity to buy directly from the growers. The ban was successful and after only a few days, baby artichokes were back in the markets.[4] From 1951–1953, the CIA study of mind-control by hypnosis and chemical intervention was known as Project ARTICHOKE. Artichokes also found themselves the butt of jokes. Curly, of US comedy trio the Three Stooges referred to an artichoke in the film Sock-a-Bye-Baby, as a barbed-wire pineapple.

Artichokes have silvery, jagged leaves and grow up to two metres tall. The edible portions of the buds consist primarily of the fleshy lower portions and the base or heart. The mass of immature florets in the middle of the bud is called the 'choke' which is inedible in older, larger flowers. Artichokes are perennials and when the terminal bud is harvested, side buds will continue to grow and these can also be eaten. The globe artichoke is not related to the Jerusalem artichoke (*Helianthus tuberosus*) which has edible tubers that may taste like globe artichokes hence taking their name.[5]

Artichokes are a good source of folic acid which helps prevent birth defects; vitamins C and K and the B vitamins; fibre, particularly in the form of inulin; minerals such as potassium, iron, copper and phosphorous and anti-oxidants such

as silymarin, ferulic acid and caffeic acid. They have among the highest anti-oxidant effects of all vegetables[6] and are commonly used in Mediterranean cuisine. Artichoke hearts top the 'spring' quarter of a 'Four-Seasons' pizza. Artichoke is used to flavour the Italian liqueur Cynar which is popular in Brazil and Switzerland as well as in Italy.

Very-long-chain inulin, extracted from globe artichokes, promotes the growth of bifidobacteria and lactobacilli and is sometimes used in prebiotics to enhance growth of 'good' bacteria in the gut.[7] Artichokes are rich in bitter-tasting polyphenolic compounds such as cynarine and chlorogenic acid, particularly in the leaves. These compounds affect the taste buds, making food and drink taste sweeter[8] and as a result, wine does not complement food containing artichokes. Chlorogenic acid and cynarine are closely related. They consist of quinic acid connected to one or two molecules of caffeic acid, respectively. Cynarine (sometimes known as cynarin) is a component of the drug Sulfad, which is used to treat liver diseases such as hepatitis and also has anti-oxidant and LDL cholesterol-reducing effects. Sulfad contains four bioactive ingredients which work synergistically. Cynarine is amphocholeretic in that it stimulates bile formation and also its elimination.

Chlorogenic acid and its analogues are also found in green coffee beans and an extract is marketed as a food supplement which reduces blood glucose levels after eating, promotes weight loss and breaks down fat reserves.[9] However, this effect is dose-dependent and at increased intake, the opposite effects have been observed.[10] Chlorogenic acids reduce blood pressure,[11] may have a laxative effect and act as a chemical sensitiser responsible for human respiratory allergy to certain plant materials.[12] The name chlorogenic comes from the Greek meaning 'giving rise to green light' reflecting the colour of the acid when oxidised.

The guaianolide sesquiterpene lactone cynaratriol is also found in artichokes. This compound has been little studied but the closely related cynaropicrin has shown anti-ageing activity by inhibiting the NF-κB activation pathway. Cynaropicrin is cytotoxic, bitter-tasting and acts as a deterrent to herbivores.[13]

The proven biological activities of compounds from artichoke support some of the traditional medicinal uses. These include use of a preparation from the leaves to treat diseases of the liver and gall-bladder, for irritable bowel syndrome (IBS) and other digestive problems, high cholesterol and blood pressure, diabetes, cancer and urinary disorders and as a cardiotonic. The effect on bile production is beneficial in many digestive, gall-bladder and liver disorders. Artichoke also mobilises stores of fat in the liver and detoxifies it, and it is a natural cholesterol-lowering agent.[14]

The stems and flower heads of the globe artichoke's cousin, the cardoon (*Cynara cardunculus*) can be eaten. Cardoon stamens contain enzymes such as cynarase which cause milk to coagulate and they are used in cheese making as a vegetarian source of rennet. The oil extracted from the seeds of cardoons is known as artichoke oil and contains the fatty acids linoleic acid, oleic acid, stearic acid and palmitic acid, in similar proportions to sunflower oil. It could potentially be used as a biodiesel. The stalks can be used in pulp and paper production.[15]

A new biorefinery in Sardinia, run by Matrica, uses artichoke oil as a feedstock. After the oil is extracted, the remaining flour is used as food for livestock. Biomass from the plant (cellulose, lignin and hemicellulose) is used to generate energy. The refinery, due to open in 2014, will initially produce azelaic acid, a starting material in the production of bioplastics which is also used in the cosmetics and pharmaceutical industries. In addition, it will generate pelargonic acid which has a potential use in phytosanitary products but mainly in the production of biolubricants. The refinery will also supply a blend of vegetable oils that can be used in the manufacture of tyres.[16]

References

1. http://rpo.library.utoronto.ca/poems/non-sum-qualis-eram-bonae-sub-regno-cynarae [Accessed 17.12.13]
2. A. Green, (2000); http://articles.philly.com/2000-06-11/food/25603016_1_artichoke-industry-artichoke-queen-castroville [Accessed 17.12.13]
3. T. Breverton, (2011); *Breverton's Complete Herbal*, Quercus Publishing Ltd, ISBN 978-0-85738-336-5, p 35
4. M.Humes, (2009); New York Times, Diner's Journal, http://dinersjournal.blogs.nytimes.com/2009/11/12/on-artichokes-and-liberty/?_r=0 [Accessed 18.12.13]
5. Mrs M. Grieve, (1973 edition); *A Modern Herbal*, Merchant Book Company Ltd., ISBN 1904779018, p 60
6. N. Ceccarelli, M. Curadi, P. Picciarelli, L. Martelloni, C. Sbrana, M. Giovannetti, (2010); *Mediterranean Journal of Nutrition and Metabolism*, 3(3), 197–201, doi: 10.1007/s12349-010-0021-z
7. A. Costabile, S. Kolida, A. Klinder, E. Gietl, M. Bäuerlein, C. Frohberg, V. Landschütze, G.R. Gibson, (2010); *British Journal of Nutrition*, 104(7), 1007–17; doi: 10.1017/S0007114510001571
8. L.M. Bartoshuk, C.-H. Lee, R. Scarpellino (1972); *Science*, 178(4064), 988–90, doi: 10.1126/science.178.4064.988
9. http://www.svetol.com/en/content/what-svetol%C2%AE [Accessed 19.12.13]
10. E. Payne, (2013); http://www.dailymail.co.uk/health/article-2332044/Is-caffeine-fix-making-fat-Study-shows-cups-coffee-day-cause-obesity.html [Accessed 19.12.13]
11. Y. Zhao, J. Wang, O. Ballevre, H. Luo, W. Zhang, (2012); *Hypertension Research*, 35(4), 370–4, doi: 10.1038/hr.2011.195
12. S.O. Freedman, R. Shulman, J. Krupey, A.H. Sehon, (1964); *Journal of Allergy*, 35(2), 97–107, doi: 10.1016/0021-8707(64)90023-1
13. F.A. Macías, A. Santana, A.G. Durán, A. Cala, J.C.G. Galindo, J.L.G. Galindo, J.M.G. Molinillo, (2012); In *Pest Management with Natural Products*, Beck, J., et al.; ACS Symposium Series; American Chemical Society: Washington, DC, Chapter 12, Guaianolides for Multipuropose Molecular Design, 167–88, doi: 10.1021/bk-2013-1141.ch012
14. http://www.rain-tree.com/artichoke.htm#.UrLiNfRdWa9 [Accessed 19.12.13]
15. J. Gominho, J. Fernandez, H. Pereira, (2001); *Industrial Crops and Products*, 13(1), 1–10, doi: 10.1016/S0926-6690(00)00044-3
16. http://www.matrica.it/default.asp?ver=en [Accessed 19.12.13]

September

11

12

13

14

15

16

17

September

18

19

20

21

22

23

24

20. Yews (*Taxus* spp.) and Paclitaxel

The European (or English) yew, *Taxus baccata*, is slow-growing and typically lives for 400–600 years. Older specimens have been estimated to be up to 5,000 years old. The famous Fortingall Yew, near Aberfeldy, Perthshire is believed to be between 2,000 and 5,000 years old. Legend has it that Pontius Pilate, the Roman Governor who oversaw the crucifixion of Jesus, was born in its shade and played under the tree as a child.[1] The tree is among the oldest living organisms in Europe. One reason for the longevity of yews is the presence of secondary shoots that grow at the base of the trunk, merging with the main trunk and giving it a ridged appearance. As the central trunk decays, these can take over the role of supporting the tree. This means that trunks are often hollow, making them more resistant to wind. Branches can also grow shoots and attach to the trunk or take root in the ground. Fallen trees can, therefore, continue to grow and trees which split under the

weight of growth can survive. The species name *baccata* means 'bearing red berries'. The open-ended berries or arils are sweet and are eaten by birds that dispel the bitter, poisonous seeds undamaged. Some birds like the hawfinch and the great tit crack open the seeds, avoiding the toxins in the seed coating, to eat the insides.

Yews were sacred to the Druids who built their temples near the trees. Early Christian churches were often located on the same sites and so the association of yew trees with churchyards began. Fronds or branches are sometimes used as a substitute for palms on Palm Sunday and Christians used to bury shoots from yew trees with the dead. The yew tree came to symbolise longevity and resurrection but it also has associations with death. The island of Iona, may have got its name from *ioua*, the Pictish word for yew. Magic wands are often made of yew wood.

The wood was also traditionally used to make longbows. A copper axe with a yew handle together with a yew longbow was discovered next to the body of Ötzi, the Iceman, the mummy found near the Italian–Austrian border, showing the use of this wood as long ago as 3,200 BC.[2] In Norse mythology, the god of the bow, Ullr, lived at Ydalir (yew dales). Bows were constructed with the heartwood on the inside and sapwood on the outside, the former resisting compression and the latter resisting stretching. However, twisted or knotted trunks could not be used therefore there was much wastage and supply could not always meet demand. Another issue was that British yew wood tended to be brittle and wood was often imported from Europe. Yew wood was also used to make lutes and other musical instruments, fence posts and canoe oars. Nowadays, it is used in veneers and by wood-turners. The trees are popular in gardens for hedging and topiary and are common bonsai trees.

The poisonous nature of the tree meant that it was not often used medicinally. More often it was used as a poison – typically when warriors wanted to commit suicide rather than be captured. However, the medieval physician Avicenna (980–1037) used a cardiac remedy, Zarnab, obtained from yew trees.[3] This has been shown to contain taxines, a group of alkaloids which act as calcium channel blockers meaning they lower blood pressure and alter heart rate.

In Japan, the Japanese yew (*Taxus cuspidata*) was used to treat diabetes and to cause abortion and the wood used to make ritual batons (*shaku*). A related species *Taxus brevifolia*, the Pacific yew, was used by Native Americans to impart strength, induce sweating and treat lung diseases and internal injuries.[4] The Pacific yew was the source of one of the best known

plant-derived medicines. In the 1950s, the US National Cancer Institute (NCI) set up the Cancer Chemotherapy National Service Center to screen compounds for anti-cancer activity. In 1960, a program of screening extracts from plants began and in 1964 an extract from the bark of the Pacific yew was found to be cytotoxic. The active compound was purified and further tested and a full report on the new compound, taxol, was published in 1971.[5] In 1979, taxol's unique method of killing cancer cells was reported. When cells divide, they are surrounded by a network of microtubules which disassemble into tubulin when the new cells are fully formed. Taxol, unlike other anti-cancer drugs such as colchicine, does not stop the microtubules forming. Instead, it stimulates microtubule formation but blocks the disassembly step so the cancer cells get clogged up and eventually die.[6]

However, there were issues with the use of taxol. The quantity of bark required to produce enough to complete testing was large – a 40 ft, 200 year old tree would only produce up to half a gram of taxol. Harvesting large quantities of bark from a tree leads to its death and it was estimated that treating a single patient would require a quantity of bark equivalent to that which would kill a tree. If taxol was to become a viable drug then other methods of production would have to be found. In 1989, with costs mounting, the NCI passed the project to the pharmaceutical company Bristol-Myers-Squibb (BMS). They were given all existing data and bark and first access to all future Federal bark supplies. In 1992, they controversially trademarked the name Taxol (now with a capital T) and the compound became otherwise known as paclitaxel.[7]

Attempts were made to extract paclitaxel from other, more abundant species or from the needles rather than the bark, which could be harvested without killing the tree. Other possibilities were to devise a synthetic method which did not require yew trees at all or a semi-synthetic route which utilised a compound with some similarity to taxol but which was easier to source. A semi-synthetic method was developed from the compound 10-deacetylbaccatin III (10-DAB) which is found in the needles of yew trees, including the abundant European yew.[8] Yields were low but another group, led by Robert Holton, improved the semi-synthesis (still using 10-DAB)[9] to the extent that the process was considered viable. BMS announced that they would no longer use Pacific yew bark after 1995. Holton also published the first total synthesis of taxol in 1994,[10] in 46 steps from the natural product patchoulene oxide which could be prepared from camphor. Other total syntheses were also developed but they shared the characteristics of many steps (30–61)[11] and low yield (0.1% in one case).

None of these methods of production proved ideal. Paclitaxel is now produced by plant cell fermentation. A *Taxus* cell line is propagated in a fermentation tank and acted on by the endophytic fungus *Penicillium raistrickii*, producing paclitaxel directly. This only requires fairly simple purification therefore represents a greener, more cost-effective alternative to synthesis or semi-synthesis. Paclitaxel can also be produced by many other endophytic fungi.[12]

Paclitaxel is used to treat ovarian, breast and lung cancers and AIDS-related Kaposi's sarcoma. Side effects can include nausea, vomiting, hair loss, infertility due to damage of the ovaries and chest pains. Some of these effects may be due to the solubiliser/emulsifier Cremophor EL (now known as Kolliphor EL) used in the formulation. As such, alternative formulations have been developed including one with paclitaxel bound to the protein albumin. Taxol was approved for use as a chemotherapy drug in 1993, almost 30 years after its activity was first noted and has become the most successful anti-cancer drug in history, treating over a million patients in the first 10 years.

Structurally-related compounds have also been isolated either from other *Taxus* species or synthetically. These are classed as taxanes and have been tested for their anti-cancer activity. A notable success is docetaxel (Taxotere) which is used to treat prostate, breast, non-small-cell lung, head, neck and ovarian cancers.[13] This can be produced semi-synthetically from 10-DAB.

References

1. http://news.bbc.co.uk/1/hi/scotland/tayside_and_central/7505706.stm
2. A. Fleckinger, (2011); *Ötzi, the Iceman The Full Facts at a Glance*, 3rd updated edition, Folio, Vienna/Bolzano and the South Tyrol Museum of Archaeology, Bolzano, ISBN 978-3-85256-574-3, p 71 and p 78
3. Y. Tekol, (2007); *Phytotherapy Research*, 21(7), 701–2, doi: 10.1002/ptr.2173
4. C.J. Earle, (2013); The Gymnosperm Database, http://www.conifers.org/ta/Taxus_brevifolia.php [Accessed 27.11.2013]
5. M.C. Wani, H.L. Taylor, M.E. Wall, P. Coggon, A.T. McPhail, (1971); *Journal of the American Chemical Society*, 93(9), 2325–7, doi: 10.1021/ja00738a045
6. P.B. Schiff, J. Fant, S.B. Horwitz, (1979); *Nature*, 277, 665–7, doi: 10.1038/277665a0
7. Opinion, *Nature*, 373, 370 (2 February 1995), doi: 10.1038/373370a0
8. J.-N. Denis, A.E. Greene, D. Guénard, F. Guéritte-Voegelein, L. Mangatal, P. Potier, (1988), *Journal of the American Chemical Society*, 110(17), 5917–9, doi: 10.1021/ja00225a063
9. R.A. Holton, (1992); *US Patent*, US 5175315 A
10. R.A. Holton, H.B. Kim, C. Somoza, F. Liang, R.J. Biediger, P.D. Boatman, M. Shindo, C.C. Smith, S. Kim, (1994); *Journal of the American Chemical Society*, 116(4),1599–600, doi: 10.1021/ja00083a067
11. BRSM Reviews, (2011); 'Taxol in 10 minutes', http://brsmblog.com/?p=929 [Accessed 29.11.2013]
12. A. Stierle, G. Strobel, D. Stierle, (1993); *Science*, 260(5105), 214–6, doi: 10.1126/science.8097061
13. S.J. Clarke, L.P. Rivory, (1999); *Clinical Pharmacokinetics*, 36(2), 99–114, doi: 10.2165/00003088-199936020-00002

September

25

26

27

28

29

30

October

1

October

2

3

4

5

6

7

8

21. Ginkgo (*Ginkgo biloba*), Ginkgolide B and Bilobalide

Ginkgo (*Ginkgo biloba*) is the last survivor of a primitive family which existed over 270 million years ago. Scientists examining the fossil record believed it was extinct until German naturalist and physician Engelbert Kaempfer (1651–1716) discovered it growing in Japan in 1691. It is native to China and seeds were taken to Japan and Korea partly because the tree is considered sacred in Buddhist temples.[1] The name ginkgo comes from the Chinese for silver apricot (*ginyo*) and biloba from the Latin, double-lobed. Leaves are unique among flowering plants, fan-like, split in the middle with veins radiating out and continually dividing into two. They resemble some segments of the leaves of maidenhair ferns (*Adiantum* spp.) and an alternative name for ginkgo is the maidenhair tree. Trees are long-lived due to their disease- and pest-resistance with examples in China claimed to be over 4,000 years old. They can form aerial roots from the branches and reproduce vegetatively from these.

More commonly, trees reproduce sexually with pollen from the cones of male trees interacting

with ovules (seeds) from female trees. Pollen grains, each containing two sperm cells, are carried by the wind to the ovules in April–May. Ginkgo trees are unusual as they, along with some other primitive plants such as the cycads, have motile sperm cells. These develop after pollination has occurred and the actual fertilisation occurs in autumn. The seeds grow as the embryo develops and this continues after the seeds fall to the ground. The outer layer of the seed (sarcotesta) contains butyric (butanoic) acid which smells like rancid butter or body odour. This has led to a preference for male trees to be grown and these are often cultivated from cuttings rather than seeds. As a result, ginkgo is classified as endangered by IUCN (International Union for Conservation of Nature) and at risk of loss of biodiversity.[1]

Leaves turn saffron yellow in autumn. Unusually, these fall in a short period of time, often 1–2 hours, thought to be governed by local climate. The fall is nicknamed ginkgo rain, occurring in the absence of wind and the day honoured as Ginkgo Day or with a Ginkgo Festival. Ginkgo is considered to be a sacred tree, symbolising longevity, a bearer of hope and a symbol of love. It is believed to offer protection against fire due to the sap secreted from the bark and also because the leaves are fire retardant. Ginkgo trees survived the great fire and earthquake in Tokyo in 1923 when other trees died and the surrounding ginkgo trees were credited with saving a temple. The ginkgo leaf was adopted as the symbol of the Tokyo prefecture. When the atomic bomb was dropped on Hiroshima in 1945, six ginkgo trees were left standing when almost all else was destroyed. Although charred, these produced healthy buds shortly afterwards and survive to this day. Elite Sumo wrestlers wear a hairstyle known as oicho-mage or big-ginkgo based on the shape of the ginkgo leaves. The whole head of hair is tied on the crown finished off with a top-knot.[1]

The bark contains groups of crystals of calcium oxalate called druse which may be present to deter herbivores. Wood is used to make chopping boards, utensils for Japanese tea ceremonies, prayer tablets and altars, insect-proof cabinets and for lacquer-ware trays and bowls. Wood from female trees can be used to make

paper. Trees are often grown as bonsai or *penjing* trees and are the national tree of China. Ginkgo leaves are used in skincare products and shampoos, as an insecticide and as a fertiliser. Dried leaves can be used in artworks or as bookmarks where their insecticidal properties protect from booklice.

The edible seed kernel is eaten in Asia. Nuts are sometimes used in congee, a type of rice-porridge and in the dish Buddha's Delight, traditionally eaten at Chinese New Year. However, over-consumption of the seeds can cause poisoning, particularly in children, due to the presence of the compound 4-methoxypyridoxine also known as ginkgotoxin which is an antivitamin B6. This causes vomiting, irritability and convulsions which can be fatal. During food shortages in Japan (1930–1960), ginkgo seeds were an important food source and many cases of toxicosis were observed, with the mortality rate reaching 27%. Ginkgotoxin is believed to decrease the amount of γ-aminobutyric acid (GABA) and to increase the amount of glutamate, causing an imbalance which can lead to convulsions. It is stable to heat so is present even after cooking or roasting of the seeds. Taking the compound pyridoxal phosphate stops the convulsions by inhibiting the action of ginkgotoxin.[2] Oil from the seeds can be used as lighting fuel and the pulp of the seeds mixed with oil or wine used as a substitute for soap. Other compounds such as bilobol, present in the leaves and seed-coatings can cause allergic skin reactions.

The seeds are commonly used in traditional Chinese and Japanese medicine to treat asthma, coughs, cancer, bladder infections and as a digestive aid. Leaves are used to treat respiratory problems, poor circulation, angina, high blood pressure, memory loss, skin diseases and as a wound plaster.[1] In Western medicine, leaf extracts have attracted most interest. These contain many compounds including the anti-oxidants quercetin and myricetin and the terpenoid lactones bilobalide and the ginkgolides. There are several related ginkgolides (including A, B, C, J, K, L, M and X). Ginkgolides (particularly B) act to block the action of platelet-activating factor so reducing blood-clotting and platelet aggregation by making platelets less sticky.[3] Platelet aggregation is linked to the development of many cardiovascular, renal, respiratory and central nervous system disorders as well as thrombosis. It is also thought to be related to the rejection of transplanted organs so ginkgolides may act to prevent this. Ginkgolide B (chemical diagram

left) can be taken to avoid reperfusion injury in organ transplantation.[4] Reperfusion injury is tissue damage caused when blood supply returns after oxygen starvation and is due to an inflammatory response and oxidative damage. Ginkgolides dilate blood vessels and leaf extracts have been used to treat pain caused by reduced blood flow to the legs (intermittent claudication). The ginkgolides are believed to be the root of the trees' longevity and also act as their defence mechanism.[1]

Leaf extracts have shown effects against Alzheimer's disease and vascular dementia, improving memory and thinking. Some studies have also indicated a preventive effect. Bilobalide (chemical diagram right) has been found to have neuro-protective properties.[5] Ginkgo is sometimes taken to aid memory in healthy people and for conditions related to reduced blood flow to the brain such as headaches and vertigo.[6] Ginkgo has shown beneficial effects against anxiety, glaucoma, macular degeneration, Raynaud's disease and pre-menstrual syndrome (PMS). Ginkgo extract is effective in reducing the clastogenic factors produced in the blood plasma after exposure to radiation such as amongst Chernobyl nuclear accident recovery workers. These factors may be responsible for the long-term damage caused by radiation exposure including cancer.[7]

Ginkgo is one of the most popular herbal supplements and is available on prescription in parts of Europe. However, recent concerns have been raised after studies showed it increased the incidence of liver cancer in mice and thyroid cancer in rats. Although there is no evidence of this effect in humans, particularly at the dosage commonly taken, some organisations in USA have now declared ginkgo unsafe. The FDA already bans its use in food or beverages and are reviewing the new evidence on herbal supplements. The issue remains contentious with advocates extolling the virtues of ginkgo supplements and dismissing the studies as using ludicrously high doses aimed at getting the supplement banned.[8]

References

1. C. Kwant, The Ginkgo Pages, http://kwanten.home.xs4all.nl/index.htm [Accessed 08.01.2013]
2. Y. Kajiyama, K. Fujii, H. Takeuchi, Y. Manabe, (2002); *Pediatrics*, 109(2), 325–7
3. X. Liu, G. Zhao, Y. Yan, L. Bao, B. Chen, R. Qi, (2012); *PLoS ONE*, 7(5), e36237, doi: 10.1371/journal.pone.0036237
4. M.L. Foegh, P.W. Ramwell, (1991); *US Patent*, US 5002965 A
5. F.V. Defeudis, (2002); *Pharmacological Research*, 46(6), 565–8, doi: 10.1016/S1043-6618(02)00233-5
6. University of Maryland Medical Center; http://umm.edu/health/medical/altmed/herb/ginkgo-biloba [Accessed 09.01.2014]
7. I. Emerit, N. Oganesian, T. Sarkisian, R. Arutyunyan, A. Pogosian, K. Asrian, A. Levy, L. Cernjavski, (1995); *Radiation Research*, 144(2), 198–205, doi: 10.2307/3579259
8. R.C. Rabin, (2013); New York Times, http://well.blogs.nytimes.com/2013/04/29/new-doubts-about-ginkgo-biloba/?_r=0

October

9

10

11

12

13

14

15

October

16

17

18

19

20

21

22

22. Cotton (*Gossypium* spp.) and Gossypol

The genus *Gossypium* is native to the Americas, Africa and India and the name cotton originates from the Arabic *al quṭn*. Plants produce yellow, cream or rose-coloured flowers which, after pollination, develop into bolls containing fibres. Fibres from the outer surface of the seed are used to make cloth. Each fibre is a long seed-coat cell, looking like a fine hair. Below the long fibres are shorter fuzzy fibres or linters.

The fibres have been used in cloth-making since at least 5,000 BC. In Northern Europe, cotton was first imported during the late medieval period but knowledge of where it came from was limited. The German name for cotton, *Baumwolle*, translates as tree wool, in keeping with the common belief as to its origin. In the mid-14th century, travel writer Sir John Mandeville related the story of a vegetable lamb native to Asia. The tree bore tiny lambs on the ends of its branches. The branches were pliable and bent down to allow the lambs to feed.[1] In fact, he was referring to cotton plants! His writings are thought to have influenced Christopher Columbus. Strange as it seems now, the existence of the so-called 'Vegetable Lamb of Tartary' was discussed by the Ancient Greeks and was still being debated by botanists in the 19th century. The vegetable lamb appears in the poem The Botanic Garden by Erasmus Darwin in 1791, written to embody the ideas of Linnaeus.

'...Rooted in earth, each cloven foot descends,
And round and round her flexile neck she bends,...

Eyes with mute tenderness her distant dam,
And seems to bleat - a vegetable lamb'[2]

By the 16th century, cotton was a valuable commodity and its use widespread. The British had plantations in the Caribbean and southern USA with a sizeable workforce largely of slave labour. Today, harvesting is carried out mainly using a machine called a cotton picker which removes just the cotton from the bolls or a cotton stripper which removes the whole bolls after application of a chemical defoliant.

After the bolls are harvested, the seed is separated from the fibres using a process called ginning. The longer fibres, staples, are twisted together to form a yarn which is spun then woven into cloth. Cotton thread is used in almost all cloth-based products including clothing, bandages, paper, coffee filters, book-binding and nappies. The linters are removed in a second ginning step and woven into a coarser fabric, lint, used in cosmetics, sausage skins, dynamite and plastics. In the UK, linters are also used to make cotton wool. Some species of cotton produce coloured fibres which could be exploited to remove the need to use synthetic dyes.[3] Cotton is a sustainable, lightweight, breathable, natural fibre.

Cotton crops are susceptible to a number of pests notably the boll weevil. Genetic modification has resulted in cotton plants containing a gene from the bacterium *Bacillus thuringiensis* (Bt) which produces a toxin harmful to many pests including bollworms but not boll weevils. Pesticides are still widely used in cotton production – while cotton is grown on 2.4% of the world's arable land, it accounts for over 24% of the world's insecticide and 11% of the world's pesticide use. However, organic cotton, which is neither genetically modified nor uses pesticides during its production, is becoming more popular.

Cotton is the most important non-food crop in the world with *Gossypium hirsutum* or upland cotton making up over 90% of commercial production. China and India are the biggest producers followed by USA. The fibres consist of over 90% cellulose, a polymeric sugar, so they have a high carbon content. Cotton efficiently converts carbon dioxide, a greenhouse gas, into carbon. Estimates suggest the annual world cotton crop removes the carbon dioxide from the atmosphere equivalent to the emissions of over 7 million cars.[4] Cellulose extracted from cotton is used to thicken ice-cream, to make chewing gum chewy and as the propellant for some fireworks. Cotton seed oil is used in soaps, margarine, cooking oil and as biodiesel.

Cotton root was used in traditional medicine to bring about abortions and was considered less powerful and safer than ergot. A preparation made from the seed was used to increase the milk of nursing mothers. The seeds were also used to treat intermittent fever (malaria). The herbaceous parts of the plant contain a large amount of mucilage and were used as a demulcent (to treat inflammation or pain in mucous membranes).[5]

Gossypol is a yellow pigment found in pigment glands in the seeds, leaves, stems and roots of the cotton plant. It interferes with

the reproduction of insects eating the plants, offering the plants some protection. Gossypol also affects reproduction in mammals and in 1929, reduced fertility in human couples who used cotton seed oil for cooking was noted. Gossypol was identified as the significant compound, affecting male fertility. This led in the 1970s, to a large scale male contraception study in China with over 8,000 men taking part, over a 10 year period. Although gossypol was found to have the desired contraceptive effect without affecting hormone levels, it had two serious side-effects. Firstly, 1–10% of patients became hypokalaemic, meaning potassium levels in the blood become too low, leading to fatigue and muscle weakness and in extreme cases paralysis.[6] However, the effect is dose-dependent and later studies suggested that hypokalaemia may not be an issue at the normal dosage used.[7] The second problem was that long-term use of gossypol could have irreversible effects on fertility. For between 5% and 25% of men, infertility remained a year after stopping taking gossypol. The effects appeared to depend on dosage and the time for which gossypol was taken – one study showed 61% regained normal fertility within 1.1 years.[8] The World Health Organisation (WHO) reviewed the data and, based on these concerns and the toxicity of gossypol (it is toxic at less than 10 times the contraceptive dose), recommended that research should be abandoned.[9] However, this was contested and research continues in China, Brazil and Africa.[6] Gossypol's sometimes irreversible effects on fertility mean it could provide a potential alternative to vasectomy.[10]

Gossypol has anti-oxidant and anti-microbial activities and lowers cholesterol levels. It is active against trypanosomes – parasites which cause diseases such as sleeping sickness and Chagas disease – and against breast, colon and prostate cancer and leukaemia.[11] Gossypol is an anti-viral agent and prevents replication of the HIV-1 virus.[12] It has activity against *Plasmodium falciparum* malaria and attempts have been made to develop less toxic (but at least equally active) derivatives. As the aldehyde (CHO) groups are the cause of the toxicity, modification of these has been carried out to produce compounds called Schiff bases in which the oxygen atom of the aldehyde is replaced by an alkylamine. These derivatives have also shown anti-malarial activity.[13]

The toxicity of gossypol also presents a barrier to the use of cotton seeds as a food source. Only ruminant animals such as cattle, with four stomachs, can tolerate gossypol. Cotton seeds contain 22% protein and use of the seeds from the commercially grown cotton crop could potentially provide the protein requirements of half a billion people. Initially mutant cotton plants which did not contain the pigment glands (glandless cotton) were developed. However, in the absence of gossypol, these were prone to attack by pests.[14]

Glandless cotton seed meal is a cost effective shrimp food. In addition, glandless cotton seed protein isolate (CPI) has shown superior foaming and emulsifying properties to soya protein isolate (SPI) and it is more soluble at acidic pH. This could, potentially, be used in sports drinks, fruit juices and tomato-based pasta sauce, whipped desserts, baked products, confectionery, ice-cream and soups.[14]

Another approach was to use a new biotechnological method known as RNA interference (RNAi). This technique was developed by Andrew Fire and Craig Mello for which they were awarded the Nobel Prize in Physiology or Medicine in 2006. Application of the method to the cotton problem by a team led by Keerti Rathore in Texas, involved 'silencing' the gene which produced gossypol in the seeds while leaving those producing gossypol in the rest of the plant unaffected. Not only have genetically modified plants been produced with gossypol levels in the seeds safe to eat but these seeds are also richer in oil than standard cotton seeds.[14] Work is continuing to assess the viability of production of cotton seeds safe for human consumption.

References

1. C.W.R.D. Moseley, (1983); *The Travels of Sir John Mandeville*, Penguin Books, ISBN-13 978-0-141-44143-6, p 165
2. E. Darwin, (1791); *The Botanic Garden, A Poem, in Two Parts*, 1825 edition, Jones & Company, London, p 143
3. L. Frost, A. Griffiths, (2001); *Plants of Eden*, Alison Hodge, ISBN 090672029X, pp 30–1
4. http://www.tjbeall.com/natural-fibers-nonwoven/ultra-clean/sustainability-and-life-cycle-info
5. Mrs M. Grieve, (1973 edition); *A Modern Herbal*, Merchant Book Company Ltd., ISBN 1904779018 p 228
6. Male Contraceptives; http://malecontraceptives.org/methods/gossypol.php#refs [Accessed 1.12.2013]
7. Z.-P. Gu, B.-Y. Mao, Y.-X. Wang, R.-A. Zhang, Y.-Z. Tan, Z.-X. Chen, L. Cao, G.-D. You, S.J. Segal, (2000); *Asian Journal of Andrology*, 2(4), 283–7
8. G.-D. Meng, Y.-Z.Hu, J.-C. Zhu, J.-H. Ding, Z.-W. Chen, X.-H. Wang, L.-T. Wong, S.-Z. Qian, G.-Y. Zhang, C. Wang, D. Machin, A. Pinol, G.M.H. Waites, (1988); *Contraception*, 37(2), 119–28, doi: 10.1016/0010-7824(88)90122-9
9. G.M.H. Waites, C. Wang, P.D. Griffin (1998); *International Journal of Andrology*, 21, 8–12, doi: 10.1046/j.1365-2605.1998.00092.x
10. E.M. Coutinho, (2002); *Contraception*, 65(4), 259–63, doi: 10.1016/0010-7824(89)90034-610.1016/S1043-4526(09)58006-0
11. X. Wang, C.P. Howell, F. Chen, J. Yin, Y. Jiang, (2009); *Advances in Food and Nutrition Research*, 58, 215–63, doi: 10.1016/S1043-4526(09)58006-0
12. B. Polsky, S.J. Segal, P.A. Baron, J.W.M. Gold, H. Ueno, D. Armstrong, (1989); *Contraception*, 39(6), 579–87, doi:10.1016/0010-7824(89)90034-6
13. V. Razakantoanina, N.K.P. Phung, G. Jaureguiberry, (2000), *Parasitology Research*, 86(8), 665–8, doi: 10.1007/PL00008549
14. C. Watkins, (2013); Inform, 24(5), 278–83, http://cottontoday.cottoninc.com/sustainability-newsroom/eating-cotton-inform_MAY_2013.pdf

October

23

24

25

26

27

28

29

October

30

31

November

1

2

3

4

5

23. Coffee (*Coffea* spp.) and Caffeine, Theobromine and Theophylline

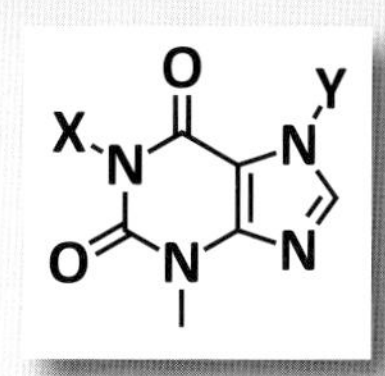

X= CH_3 Y= CH_3 **Caffeine**
X= H Y= CH_3 **Theobromine**
X= CH_3 Y= H **Theophylline**

Coffee is one of the world's most important crops, second only to petroleum in global trading. Over 400 billion cups are consumed per year and it is the world's second most popular drink after water. Its origins may lie in Ethiopia where the story has Kaldi, a goat herd, finding his unusually frisky flock feeding on the red berries of coffee plants. He tried the berries himself and shared the harvest with a nearby monastery. The monks were not impressed, throwing the berries in the fire, but the aroma of roasting beans won coffee a reprieve and the monks cultivated a coffee bush, brewing a drink from the berries to help them stay awake during prayers. Other legends attribute the origins of coffee to Yemen, particularly the city of Mocha. Coffee was mainly chewed in balls with animal fat. It was also brewed into a broth in Arabia and the Arabians protected their dominance of production by boiling beans for export, making them infertile.[1]

Marco Polo (*ca.* 1254–1324) is credited with introducing coffee to the West when he brought it back to Venice in 1295. However, coffee didn't become popular in Europe until the 1600s after an Arabian trader, Baba Budan smuggled out fertile coffee seeds and these reached European colonies. The Dutch were the first Europeans to grow coffee, in their colony of Java. A cutting from a coffee tree gifted by

the Dutch to the French was taken to French Martinique by a naval officer, Gabriel Mathieu de Clieu. During the trip, the ship survived a storm, an attack by pirates and the coffee seedling was attacked by a mad passenger. Water was rationed and de Clieu shared his rations with the seedling until they safely reached land. Under armed guard, the seedling flourished and over the next 50 years its family grew to 18 million bushes, spreading throughout Latin America. Brazil wanted to join the lucrative coffee market and in 1727, Lt. Col. Francisco de Melo Palheta was sent on a visit to French Guiana with the pretence of negotiating over a border dispute. In fact, Palheta's aim was to charm the French Governor's wife into giving him a bouquet spiked with coffee seedlings. He was successful and Brazil is now the largest producer of coffee, growing about a third of the world's supply.[1]

Coffee houses became fashionable meeting places for conducting business in the 17th and 18th centuries and both Lloyd's of London and the London Stock Exchange originated there. However, Nicholas Culpeper (1616–54), the apothecary and physician, was a scathing critic, writing 'the coffee-liquor… stinks most loathsomely… the proponents of this filthy drink affirm that it causes watchfulness… they do also say that it makes them sober when they are drunk'. He concluded that 'neither can it be endowed with any such properties as the indulgers of it feed their fancies with'.[2]

The most common species of coffee is *Coffea arabica*, accounting for 70–80% of the crop. Trees or bushes produce red or purple fruits called cherries usually containing two seeds which are known erroneously as beans. They grow with a flat side facing each other. In about 5–10% of cherries, only one seed is fertilised and this is smaller and rounder. These are known as peaberries and are separated from the main crop. Traditionally, cherries are picked by hand. However, they ripen on a tree at different times thus harvest is labour-intensive and mechanised picking is becoming more common. The beans are removed from the cherries and processed before roasting.

For one speciality coffee, beans are fed to Asian palm civits. These are collected after they have passed through the digestive tract and used to make the drink Kopi Luwak, commonly known as 'crap coffee'. This is prized for its low bitterness and as the popularity (and price) has increased, concerns have been raised about the conditions endured by the civits.[3] A cup of Kopi Luwak can fetch £70 in London. Instant coffee offers a much more simple method of producing a drink from the beans. It was invented in 1906 by George C. Washington, an English-Belgian chemist. He experimented on the powdery residue left after brewing coffee and developed a method of drying to allow coffee to be made simply by adding water.

Coffee is best known as a stimulant. This action is due to bitter-tasting alkaloids, most significantly caffeine, which protect the plant from predators. To reduce the stimulant action, coffee can be decaffeinated. Caffeine is removed from the beans either by the chemical-free Swiss Water process or by steaming followed by removal of the caffeine-containing oils using solvents. The crude caffeine by-product is purified and sold for use in the beverage or pharmaceutical industries. Recent research has suggested that the crude caffeine has medicinal benefits as a neuro-protective agent and has anti-oxidant, anti-inflammatory and anti-hyperglycaemic effects.[4]

Caffeine is present in cola nuts, tea and coffee, acting as a natural pesticide against some insects. In humans, it has been ascribed a variety of health effects both positive and negative. Long-term consumption has been linked with reduced risk of cardiovascular disease, Parkinson's disease and type 2 diabetes. By blocking adenosine receptors, it acts as a stimulant to the central nervous system. Acetylcholine is released which may have positive effects on memory and cognition[5] and may reduce age-related mental decline. Caffeine reduces fatigue and increases alertness and vigilance. This has led to the popularity of high-caffeine drinks and supplements with sometimes fatal consequences. The consumption of numerous caffeine mints was blamed for the death of John Jackson in 2013. Each mint contains 80 mg of caffeine, the equivalent to a can of energy drink.[6] Concerns have also been raised over excessive consumption of energy drinks leading to their ban in France in 2004.[7]

Caffeine is a diuretic and vasoconstrictor and is used to treat migraines. However, it raises blood pressure and it can also cause decreased control of fine motor movements (e.g. causing shaky hands). Pregnant women are advised to limit their consumption to minimise the risk of miscarriage and still birth. In large amounts, caffeine can cause anxiety and insomnia.[5] As caffeine is considered a psychoactive drug, it is banned by practitioners of some religions.

A recent study showed it is not just humans who get a buzz from caffeine. The nectar from coffee plants contains caffeine in quantities small enough not to adversely affect its taste. Bees feeding on caffeine-containing nectar were three times more likely to revisit the same plants after 24 hours and still twice as likely after 72 hours. The authors conclude caffeine may improve the bees' foraging ability while benefiting the plants with a more faithful pollinator.[8] Caffeine, however, has a pronounced detrimental effect on the ability of house spiders to make webs.[9]

Two other structurally related compounds, theobromine and theophylline are metabolites of caffeine in the liver. Theobromine also occurs in tea, cola nuts and mainly, in cocoa, obtaining its name from the plant *Theobroma cacao*. This genus name derives from the Greek for 'food of the gods'. Like caffeine, theobromine is a diuretic and stimulant but it is only slowly metabolised by some animals. Theobromine in chocolate renders it toxic to dogs[10] and cats but as cats can't taste sweetness, cases of toxicosis in cats are rare.[11] Chocolate's reputed aphrodisiac effects are partly due to theobromine. It is a vasodilator and formation of theobromine in the body from caffeine has led to caffeine being referred to, confusingly, both as a vasodilator and a vasoconstrictor. While caffeine itself is a vasoconstrictor, it is the metabolite theobromine which causes vasodilation. The combination of diuretic and vasodilation properties means theobromine is a potential treatment for hypertension. It has a smaller effect on the central nervous system but a larger stimulant effect on the heart than caffeine. Theobromine can be used against persistent cough, acting by suppressing misfiring of the vagus nerve[12] and also to treat asthma by relaxing the smooth muscles.

Theophylline can also be used to treat respiratory diseases such as chronic obstructive pulmonary disease (COPD) and asthma. However, it has serious interactions with other drugs such as phenytoin and has a low therapeutic index therefore is now rarely used. Theophylline has also been shown to improve poor sense of smell in patients where depleted levels of two proteins cAMP and cGMP are implicated.[13] It is found primarily in cocoa beans and tea. The major caffeine metabolite in the liver is another related alkaloid, paraxanthine. This makes up 84% of the metabolite mixture compared to 12% theobromine and 4% theophylline. However, unlike the other two, paraxanthine does not occur in plants and is present only in animals who have ingested caffeine.

References

1. National Geographic; http://www.nationalgeographic.com/coffee/ax/frame.html [Accessed 03.12.2013]
2. T. Breverton, (2011); *Breverton's Complete Herbal*, Quercus Publishing Ltd, ISBN 978-0-85738-336-5, p 145
3. O. Milman, (2012); The Guardian, http://www.theguardian.com/environment/2012/nov/19/civet-coffee-abuse-campaigners [Accessed 03.12.2013]
4. Y.-F. Chu, Y. Chen, P.H. Brown, B.J. Lyle, R.M. Black, I.H. Cheng, B. Ou, R.L. Prior, (2012); *Food Chemistry*, 131(2), 564–8, doi: 10.1016/j.foodchem.2011.09.024
5. R. Thompson, K. Keene, (2004); *The Psychologist*, 17(12), 698–701
6. P. Cheston, L. Smith, (2013); http://www.independent.co.uk/news/uk/home-news/man-died-after-overdosing-on-caffeine-mints-8874964.html [Accessed 03.12.2013]
7. C. Nordqvist, (2004); http://www.medicalnewstoday.com/releases/5753.php [Accessed 03.12.2013]
8. G.A. Wright, D.D. Baker, M.J. Palmer, D. Stabler, J.A. Mustard, E.F. Power, A.M. Borland, P.C. Stevenson, (2013); *Science*, 339(6124), 1202–4, doi: 10.1126/science.1228806
9. D.A. Noever, R.J. Cronise, R.A. Relwani, (1995); NASA Tech Briefs 19(4), 82. Published in New Scientist, (1995), issue 1975, http://www.newscientist.com/article/mg14619750.500-spiders-on-speed-get-weaving.html
10. http://www.vetrica.com/care/dog/chocolate.shtml [Accessed 03.12.2013]
11. D. Biello, (2007); http://www.scientificamerican.com/article.cfm?id=strange-but-true-cats-cannot-taste-sweets [Accessed 03.12.2013]
12. http://www.bbc.co.uk/news/health-12048275 [Accessed 03.12.2013]
13. http://www.sciencedaily.com/releases/2008/04/080407114619.htm [Accessed 03.12.2013]

November

6

7

8

9

10

11

12

November

13

14

15

16

17

18

19

24. Aloes (*Aloe* spp.) and Aloin B

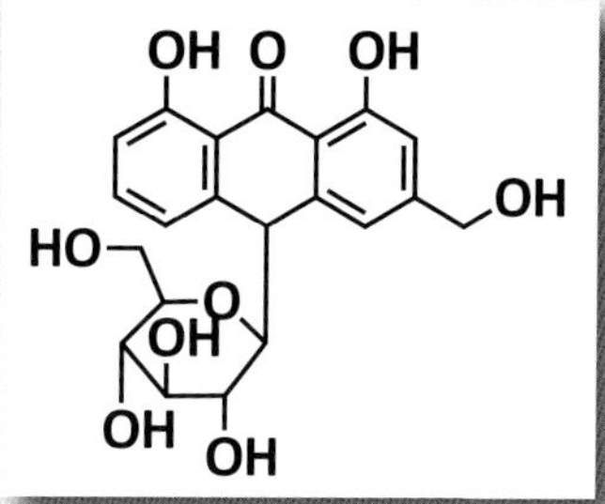

The genus *Aloe* consists of over 500 succulent species native to Africa. The fleshy leaves have serrated edges and store water, allowing survival in arid conditions. The most common species, *Aloe vera,* is cultivated for use in the cosmetics, food, supplement and pharmaceutical industries and as an ornamental plant. Common names include true aloe (*vera* means true or genuine), first aid plant and medicine plant. Other important species include *Aloe ferox* (Cape aloe or bitter aloe) and *Aloe perryi* (Socotrine aloe). The name Socotrine relates to a story of Alexander the Great. He was wounded during the siege of Gaza in 330 BC and the wound became infected as he travelled through North Africa. A priest, recommended by his tutor Aristotle, treated the infected wound with aloes from the nearby island of Socotra. The wound healed miraculously and Aristotle persuaded Alexander to invade Socotra to secure a supply of aloes for his army. Socotrine aloe leaves can also be used to make a purple dye which requires no fixative.[1]

Included in the genus *Aloe* are some trees such as *Aloe barberae* (tree aloe) and *Aloe dichotoma.* The latter species, native to the Northern Cape region of South Africa, is commonly known as the quiver tree. This name relates to the reputed use of hollowed out branches to form quivers for their arrows, by the indigenous people.

it has been suggested that aloes were used to treat Jesus after crucifixion.[2] Confusingly, some references to aloes in the Bible and other ancient texts probably refer to *Lignum aloes*, from the tree *Aquilaria malaccensis*, which were used as incense.[1] Aloes were associated with long life and a drink made from aloes, the 'Elixir of Jerusalem', was taken by the Knights Templar during the Crusades. In Russia, aloe juice was termed the 'Elixir of Longevity' and the Greeks associated aloe with fortune, beauty and good health. Cleopatra is believed to have bathed in aloe juice before meeting with Mark Antony. It was believed to ward off evil spirits and to symbolise patience. Medicinally, it was used to treat digestive disorders and skin conditions, to promote hair growth and as a laxative. Aloes are now used throughout the world and there is a Museum of Aloe (Museo de Aloe de Lanzarote) in the village of Arrieta in Lanzarote.

Aloes contain two important components – the gel and the latex. The gel is a clear, jelly-like substance found in the inner part of the leaves which solidifies on contact with air. Diluted gel is known as juice. The yellow, sticky latex is found just under the surface of the leaves. Some products are made from crushed leaves so contain both gel and latex, but as the medicinal properties of each are quite different, these are usually separated and may be processed to remove any traces of the other. Aloes can cause uterine contractions and should be avoided during pregnancy and when breast-feeding.

Powdered aloes may have been used in the embalming process of pharaohs in ancient Egypt, although aloes could also be present in tombs for medicinal or spiritual use. Aloes and myrrh were found in the tomb of Jesus and

The latex contains the aloins, aloin A and aloin B (also known as barbaloin and isobarbaloin). These are anthraquinone pigments and differ only in the relative orientation of the constituent sugar group. The presence of this sugar group means they are termed glycosides. They are yellow-coloured and bitter-tasting. The quantities present and ratio of aloins A and B vary with *Aloe* species. Socotrine aloe contains no, or almost no, aloin-B.[1] Aloins have laxative effects, increasing the peristaltic contractions in the colon. In the gut, they decompose and one decomposition product, aloe-emodin-9-anthrone, has been shown to increase water content in the intestines and also to enhance mucous production.[3] They were formerly used in over-the-counter laxatives but their use was banned in 2002 in USA when sufficient safety data could not be provided.[4] However, they are still licensed for use in Germany and are used in veterinary medicine.

Aloins are used in small quantities as bittering agents particularly in alcoholic beverages where they are listed as natural flavour. Aloins cause melanin aggregation, leading to skin lightening and therefore have a potential use in cases of hyperpigmentation.[5] They also have anti-cancer properties against melanoma which are improved when used in combination with the compound cis-platin.[6] Aloin A is used as a starting material in the synthesis of diacerein, a drug used in the treatment of osteoarthritis. The aloins may be the cause of the carcinogenic effects observed in some animal tests on aloes.[7] These were conducted using 'whole leaf extract' (known as non-decolourised) therefore contained both latex and gel. The latex also contains aloe emodin which is similar in structure to aloin but without the sugar group. It has a strong purgative effect and is also found in cascara (*Rhamnus purshiana*) and the leaves of *Senna alexandrina*, both of which are used as natural laxatives. It has a marked anti-viral effect against herpes simplex viruses.[8]

Aloe gel contains little or no aloins and is used as a digestive healer and, externally, to treat skin infections or burns and promote wound healing. The evidence of its medicinal effects is mixed but some benefits have been proven. The gel can be used to treat ulcerative colitis, psoriasis, some cases of dermatitis, dry skin, mild to moderate burns, lichen planus infections, UV-induced erythema (i.e. sunburn) and genital herpes. It raises the pH in the stomach so can be beneficial in cases of acid reflux. The gel also reduces cholesterol and blood glucose levels and can be used to treat angina pectoris and kidney stones.[9] One of the chemical components of the gel, acemannan, is a polymeric sugar which shows anti-inflammatory effects. This has been shown to be useful in treating alveolar osteitis (dry socket i.e. inflammation of the bone exposed after tooth extraction).[9] Acemannan also has anti-viral and anti-neoplastic activities and affects the gastrointestinal system.[10] The potential use of aloe to treat radiation burns led to aloes or aloe gel products being stockpiled by the USA during the Cold War.[11] Gel has also been used to facilitate artificial insemination of sheep.[12] In the food industry, it can aid the retention of desirable compounds and retard ripening when used as a coating for table grapes.[13]

When aloe gel is exposed to air, it is oxidised and loses some of its beneficial properties. However, the gel can be stabilised thus preserving the properties of fresh aloe. Aloe gel has many other uses. Traditionally, huntsmen in the Congo covered their bodies with the gel to reduce perspiration which would alert potential prey to their presence. In India it is used to treat asthma and in parts of Africa, to treat conjunctivitis. The gel is also used against hypertension.[9] Due to its moisturising and skin softening properties, *Aloe vera* gel or its derivatives is widely used in the cosmetics industry in moisturisers, shaving gels, suncreams, soaps, tissues and make-up.

Aloe juice can contain anything from 15% to over 85% gel. It is promoted as a health drink containing a range of vitamins and minerals and is one of the few plant based sources of vitamin B12. It is also claimed to rid the body of toxins.[14] Aloe has been used to make a toothpaste that may be beneficial for sensitive teeth and gums irritated by conventional pastes. *Aloe vera* seed oil has been suggested as a possible biodiesel.[15]

References

1. Mrs M. Grieve, (1973 edition); *A Modern Herbal*, Merchant Book Company Ltd., ISBN 1904779018, pp 26–9
2. www.tombofjesus.com/index.php/en/component/content/article/53-crucifixion/events/76-aloe-and-myrrh [Accessed 07.12.2013]
3. Aloin A: Y. Ishii, H. Tanizawa, Y. Takino, (1994); *Biological and Pharmaceutical Bulletin*, 17(5), 651–3

   and Aloin B: Y. Ishii, Y. Takino, T. Toyo'oka, H. Tanizawa, (1998); *Biological and Pharmaceutical Bulletin*, 21(11), 1226–7
4. M.M. Dotzel, (2002); Food and Drug Administration, Docket No. 78N-036L, http://www.fda.gov/ohrms/dockets/98fr/050902a.htm
5. S.A. Ali, J.M Galgut, R.K. Choudhary, (2012); *Planta Medica*, 78(8), 767–71, doi: 10.1055/s-0031-1298406
6. C. Tabolacci, S. Rossi, A. Lentini, B. Provenzano, L. Turcano, F. Facchiano, S. Beninati, (2013), *Amino Acids*, 44(1), 293–300, doi: 10.1007/s00726-011-1166-x
7. A.R. Pandiri, R.C. Sills, M.J. Hoenerhoff, S.D. Peddada, T.-V.T. Ton, H.-H.L. Hong, G.P. Flake, D.E. Malarkey, G.R. Olson, I.P. Pogribny, N.J. Walker, M.D. Boudreau, (2011); *Toxicologic Pathology*, 39(7), 1065–74 , doi: 10.1177/0192623311422081
8. D.G. Owen, R.J. Sydiskis, (1987), *US Patent*, US 4670265 A
9. O. Grundmann, (2012); *Natural Medicine Journal*, http://www.naturalmedicinejournal.com/article_content.asp?article=356
10. http://pubchem.ncbi.nlm.nih.gov/summary/summary.cgi?sid=596005&loc=es_rss [Accessed 07.12.2013]
11. http://www.aloe1.com/admiral-frank-voris-md/ [Accessed 07.12.2013]
12. F. Rodriguez, H. Baldassarre, J. Simonetti, F. Aste, J.L. Ruttle, (1988); *Theriogenology*, 30(5), 843–54, doi: 10.1016/S0093-691X(88)80046-3
13. M. Serrano, J.M. Valverde, F. Guillén, S. Castillo, D. Martínez-Romero, D. Valero, (2006); *Journal of Agricultural and Food Chemistry*, 54(11), 3882–6, doi: 10.1021/jf060168p
14. B. London, (2013); http://www.dailymail.co.uk/femail/article-2431575/Move-coconut-water-Aloe-Vera-juice-tipped-seasons-hottest-health-drink.html [Accessed 07.12.2013]
15. P.S. Rathore, P. Mangalorkar, P.S. Nagar, M. Daniel, S. Thakore, (2013); Energy & Fuels, 27(5), 2776–82, doi: 10.1021/ef301950j

November

20

21

22

23

24

25

26

November

27

28

29

30

December

1

2

3

25. Prickly Pears (*Opuntia* spp.) and Quercetin

Opuntia is a genus of the cactus family containing over 200 species named after the ancient Greek city of Opus. They are also known as paddle cacti or nopales or, commonly, as prickly pears. The name nopales is from the Nahuatl (a Mexican language) word for the pads. Over 100 species are native to Mexico where the pads are eaten raw or cooked and described as tasting like green beans. The coat of arms of Mexico has an eagle perched on an *Opuntia* cactus eating a rattlesnake. The young pads can be peeled to remove the spines which are not fully hardened. The fleshy parts are eaten as a vegetable known as nopalito. Nopales can also be candied and are then known as acitróne. Nopales have a low glycaemic index so stabilise blood sugar levels.[1] The most commonly cultivated species is *Opuntia ficus-indica* (Indian fig opuntia).

The pads are flattened stems known as cladodes.They usually have two types of spines – large, smooth, fixed spines and smaller hair-like spines, called glochids. The smooth spines were used to play 78 rpm records which were made of shellac until the 1950s. Shellac is a resin from the female lac bug, found in certain trees in India and Thailand. The resin is hard but damaged easily by steel styli, therefore even today, collectors avoid using them. The softer natural styli made from bamboo or prickly pear spines, cause less damage to the records, although they do get worn themselves. Glochids easily detach from the plant and embed in the skin of predators. The shafts are barbed so difficult to remove. Some species in the genus, such as the beavertail cactus (*Opuntia basilaris*) or the bunny ears (or polka-dot) cactus (*Opuntia microdasys*) have no spines. For protection, they have clusters of glochids on the pads.

As well as the pads, the fruits, sometimes known as tuna, can also be eaten and these are popular across southern Europe and in parts of South America, Asia and Africa. The spread of *Opuntia* to Europe may have been due to fruits being carried on ships for the crew to eat in order to avoid scurvy. The outer skin is removed before eating and the fruit is often chilled. The fruit contains numerous small seeds and tastes like watermelon. Fermented fruits are used to produce a mildly alcoholic drink called colonche, popular in Mexico.

The flesh, or mucilage, contains exopolysaccharides – high weight molecules made up of sugars which function as part of the plant's water storage system. Traditionally, this was used in Mexico to purify water and recent research has shown that the mucilage acts as a flocculent. This attaches to sediment and bacteria in the water causing it to drop to the bottom of the vessel. Tests show 98% of bacteria can be removed using this method.[2] Functional groups in the mucilage (carboxyl, carbonyl and hydroxy) also interact with problematic heavy metals such as arsenic, allowing them to be removed from the water.[3] The mucilage could have a further application as an oil dispersant and absorbent, causing oil globules to break up and disperse more quickly. This could provide a green alternative to the chemical dispersants currently used.[4]

Prickly pears can also absorb selenium from the soil thus improving the soil for other uses.[5] The edible parts of the plant are a good source of selenium in the diet. Selenium improves the elasticity of the skin, is an anti-oxidant and may be beneficial to prostate health. *Opuntia* species contain some psychoactive compounds such as mescaline and have been added to the entheogenic drink ayahuasca. This is used by some shamans as a tool of enlightenment. Flowers are astringent and used to reduce bleeding and treat disorders of the gastrointestinal tract. The flowers and stems have anti-spasmodic, emollient and diuretic properties. The juice can be used as a waterproof binding agent in earthen plaster and in adobe, a natural building material made of sand, clay and water. Gum from the stems can be mixed with oil to make candles.[6] In traditional medicine, prickly pears were used to treat diabetes and high cholesterol. Both of these uses have been clinically validated.[1,7]

Prickly pears contain the red pigment betanin, also known as beet red and found in beetroot. However, the major contribution of the genus to the dyeing industry is in the production of cochineal. The sessile (i.e. immobile) parasites *Dactylopius coccus* live on the moisture and nutrients from host *Opuntia* plants and produce carminic acid to deter predators. This is extracted and mixed with aluminium or calcium salts to produce carmine dye which is also known as cochineal. It was used as a dye by the Aztecs and Mayans and is now used as a food colouring and in cosmetics. Nopale farms (nopalries) for the production of cochineal are mainly located in Peru, the Canary Islands and Chile.

Cattle ranchers in southern USA sometimes use prickly pears both as a source of cattle-feed, particularly during times of drought, and also as a boundary fence. Spines are burned off to prevent damage to the feeding cattle. The high water content reduces the requirement for water from other sources.[8] In 1961, Cuban president Fidel Castro's troops planted an 8-mile-long barrier of the plants around the north-eastern section of the fence around the US base at Guantanamo Bay, to prevent Cubans approaching the base to seek refuge. This became known as the Cactus Curtain, an allusion to the Iron Curtain.[9]

The hazards of the plants, particularly when picking the fruit, are highlighted in the song 'The Bare Necessities' from Disney's *The Jungle Book*.

'Now when you pick a pawpaw
Or a prickly pear
And you prick a raw paw
Next time beware
Don't pick the prickly pear by the paw
When you pick a pear
Try to use the claw'[10]

There are other problems to bear in mind when eating the fruit. *Opuntia ficus-indica* fruit can cause constipation when eaten with the seeds but have a laxative effect when eaten without.[11]

Prickly pears also contain the flavonol quercetin, a yellow/red pigment found widely in fruit and vegetables, which has a number of health benefits. Good sources are capers, onions, green tea, apples and berries where quercetin is mainly found in the skin. There is also quercetin in red wine, a result of its presence in the skins of red grapes. Quercetin is an anti-oxidant and anti-inflammatory and can be used to treat high cholesterol, heart disease, high blood pressure and circulation problems. It protects low-density lipoprotein (LDL) cholesterol from oxidative damage caused by free radicals. The damaged cholesterol would be deposited in and damage blood vessels, therefore prevention of the free radical attack mediates the detrimental effects of LDL cholesterol.[12]

Quercetin is sold as a food supplement and has been suggested for the treatment of a number of conditions. These include diabetes, obesity,[13] cataracts, hay fever and other allergies, asthma, gout, viral infections, chronic fatigue syndrome and chronic infections of the prostate or bladder (interstitial cystitis). It may also inhibit tumour growth in some cancers.[14] Quercetin could, potentially, be used to increase endurance and improve athletic performance.[15] It causes the number of mitochondria in muscle cells to increase thus improving energy production.

References

1. M. Bacardi-Gascon, D. Dueñas-Mena, A. Jimenez-Cruz, (2007); *Diabetes Care*, 30(5), 1264–5; doi:10.2337/dc06-2506
2. A.L. Buttice, J.M. Stroot, D.V. Lim, P.G. Stroot, N.A. Alcantar, (2010); *Environmental Science & Technology*, 44(9), 3514–9, doi: 10.1021/es9030744
3. D.I. Fox, T. Pichler, D.H. Yeh, N.A. Alcantar, (2012); *Environmental Science & Technology*, 46(8), 4553–9, doi: 10.1021/es2021999
4. N.A. Alcantar, D.I. Fox, S. Thomas, (2013); *U.S. Patent*, US 20130087507 A1
5. D. O'Brien, (2012); *Agricultural Research*, 60(1), 12–13, http://www.ars.usda.gov/is/AR/archive/jan12/cactus0112.htm [Accessed 09.12.2013]
6. L. Frost, A. Griffiths, (2001); *Plants of Eden*, Alison Hodge, ISBN 090672029X, pp 40–1
7. M.A. Gutierrez, (1998), *Nutrition Bytes*, 4(2), http://escholarship.org/uc/item/2x53d917
8. J.C. Paschal; http://agrilife.org/coastalbend/nutritional-value-and-use-of-prickly-pear-for-beef-cattle/ [Accessed 09.12.2013]
9. Time Magazine, (1962); http://content.time.com/time/magazine/article/0,9171,940656,00.html [Accessed 09.12.2013]
10. Lyrics by Terry Gilkyson; http://disney.wikia.com/wiki/The_Bare_Necessities [Accessed 09.12.2013]
11. F. Lentini, F. Venza, (2007); *Journal of Ethnobiology and Ethnomedicine*, 3:15, doi: 10.1186/1746-4269-3-15
12. T. Hayek, B. Fuhrman, J. Vaya, M. Rosenblat, P. Belinky, R. Coleman, A. Elis, M. Aviram, (1997); *Arteriosclerosis, Thrombosis, and Vascular Biology*, 17, 2744–52, doi: 10.1161/01.ATV.17.11.2744
13. L. Aguirre, N. Arias, M.T. Macarulla, A. Gracia, M.P. Portillo, (2011); *The Open Nutraceuticals Journal*, 4, 189–98
14. D.M. Lamson, M.S. Brignall, (2000); *Alternative Medicine Review*, 5(3), 196–208
15. University of Maryland Medical Center; http://umm.edu/health/medical/altmed/supplement/quercetin [Accessed 09.12.2013]

December

4

5

6

7

8

9

10

December

11

12

13

14

15

16

17

26. Birch (*Betula* spp.) and Betulinic Acid

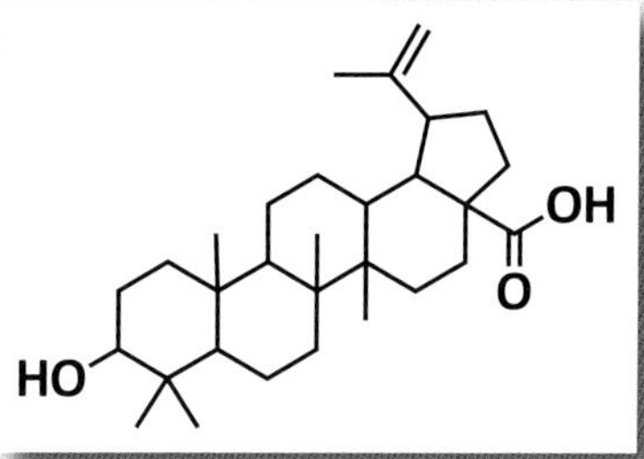

The most common members of the genus *Betula*, the birches, found in Britain are the silver birch (*Betula pendula*) and the white or downy birch (*Betula pubescens*). The bark has large lenticels which function as pores to allow gases in and out. They can be diamond- or eye-shaped, earning birch its nickname 'the watchful tree'. In some species, papery sections of the bark peel off and this has been used for writing on. The oldest surviving handwritten documents found in Britain, the Vindolanda tablets from 1st and 2nd century AD, were written on wood bark, including birch. The name birch may derive from the Sanskrit *bhurga* meaning 'a tree whose bark is used for writing upon'.

Birch wood is light yet strong and is used to make bobbins, spools and reels. In 1942, the US military needed new transport vehicles that could carry heavy loads but due to wartime shortages these could not be made of aluminium. Hughes Aircraft Company, owned by Howard Hughes, developed a flying boat made almost entirely of birch wood, the Hughes H-4 Hercules. Although made mainly of birch, it acquired the nickname 'The Spruce Goose'. Unfortunately, the plane was not completed until 1947, by which time the need for it had passed. Hughes was called to testify before a congressional committee to justify the 22 million dollars spent by the US government on the project. During the hearing, Hughes himself piloted the plane on its one and only flight. The

plane was kept ready to fly until Hughes' death in 1976 and is now housed in the Evergreen Aviation Museum in Oregon.[1]

Baltic or Russian birch is used to make speaker cabinets, drums and as a tonewood to make guitar bodies. The sap can be extracted from birch trees by drilling a hole in the trunk and it can be drunk or used as a flavouring. Fermented, it produces birch wine[2] and the oil distilled from the sap is used to make birch beer, a soft drink like root beer. The sap contains sugars including fructose, glucose and xylitol and is concentrated to form a syrup used as a tonic in some counties. In Russian folk medicine, birch syrup is used as an antiseptic, anti-parasitic, anti-inflammatory and anti-itching agent. An infusion made from the twigs and sap was used as a shampoo and birch sap is still used in some hair care products. Sweet birch (*Betula lenta*) was formerly used in the commercial production of oil of wintergreen (methyl salicylate) which is used as a rubefacient, antiseptic, flavouring and fragrance. This compound is now produced synthetically.

Birch oil is used to saturate the tanned hide in the production of Russian leather. This makes it waterproof, acts as an insect repellent and gives its characteristic smell. This smell was key in 1768, when the Russian Count Orlof ordered a perfume to be developed by perfumers Bayleys of London, to evoke the aroma of the Russian Court. The resulting perfume, Eau de Cologne Imperiale Russe was used to scent Imperiale Russian Leather soap which later became known as Imperial Leather.[3]

Umeå in Sweden, where novelist Stieg Larsson grew up, is sometimes known as the city of birches. In 1888, a fire raging in the city was supposedly halted by birch trees and afterwards 3,000 silver birches were planted as fire protection. In Scandinavia the inner bark of silver birch trees (phloem) was traditionally ground to make bark bread in times of hardship. Carl Linnaeus expressed concern that the forests were being devastated by the peasants harvesting bark to make flour.[4]

A bound bundle of sticks, typically of birch, tied to an axe is known as a fasces. In ancient Rome, this was the symbol of the magistrates, denoting strength through unity. It was adopted by Benito Mussolini as the symbol of his Italian Fascist Party, which also got its name from the Italian word *fascio*, for fasces. The symbol is still used widely in the USA and elsewhere.

Birch twigs are used to make besoms (brooms) with ash or hazel handles. As well as their use for sweeping, they are also used in Wicca (modern witchcraft). Twigs are used in Scandinavian saunas as a tool for self-massage and to increase circulation and open the pores. The twigs are usually soaked in water prior to use to soften them. Traditionally, a bundle of birch twigs with the leaves removed was used to beat offenders (birching) as a form of corporal punishment. In the UK, birching was used from the mid-19th century until it was abolished in 1948 (1993 in the Isle of Man).[5] Twigs are also used to make jumps for horses and leaves and bark used to make yellow or brown/pink dyes.

Silver birch trees are effective at reducing air-borne pollution from particulates. An experiment tested the effects of a row of silver birch trees on the pavement in front of houses on the busy A9 in Lancaster. After 2 weeks, the pollution levels within houses with silver birch trees outside were 50–60% less than in those houses without them. This is thought to be due to small hairs and ridges on the leaves trapping the pollution particles. As the leaves are sparse, filtering is more effective than by more densely-leaved trees. Particles are washed off by rain thus allowing filtering to continue.[6]

Birch tree pollen is one of the most important causes of hay fever symptoms in northern latitudes. It contains the protein Bet v 1 which is the significant allergenic factor. Birch has many medicinal uses in herbal medicine. An infusion of the leaves can be used to treat gout, dropsy, rheumatism, cystitis, kidney stones and the sap and leaves are diuretic. The oil is astringent and used for eczema while the inner bark is used to treat intermittent fevers. The young shoots and leaves excrete an acidic resin which can be combined with alkalis and used as a tonic laxative.[7] The bark can be applied externally, with the wet internal side against the skin, to ease muscle pain.[8] The beneficial effect may be due to the presence of salicylates, as in willow.

White-coloured birch barks contain the compounds betulinic acid and its precursor betulin. As well as giving the bark its characteristic colour, these compounds may protect the tree against extremes of temperature, parasitic infections and solar radiation. Both of these compounds have extensive medicinal effects, with betulinic acid generally the more active of the two. Betulinic acid has activity against several cancers. It induces apoptotic (programmed) cell death in cancer cells leaving normal cells unaffected.[9] It inhibits the enzyme topoisomerase which regulates the winding of DNA strands, causing breaks in the DNA which lead to cell death. Betulinic acid also has anti-inflammatory, anti-malarial and anti-retroviral (anti-HIV-1) effects.[10]

Betulin was isolated from birch by the German-Russian chemist, Johann Tobias Lowitz (1757–1804) in 1788. This was one of the first active compounds isolated from a plant. Lowitz also developed the use of charcoal as a purifying agent for drinking water. An emulsion of betulin can be used to treat neuro-dermatitis and psoriasis. Betulin has been shown to decrease the biosynthesis of cholesterol and fatty acids, increase insulin sensitivity and reduce atherosclerotic plaques, suggesting possible uses in the treatment of high cholesterol, atherosclerosis and type 2 diabetes.[11] It is also active against the herpes virus.

An issue with both betulin and betulinic acid is their poor solubility which limits the

Chaga mushroom (*Inonotus obliquus*) and chaga tea

bioavailability in the body. The parasitic chaga mushroom (*Inonotus obliquus*) grows predominantly on birches and converts betulin to a more bioavailable form of betulinic acid. The mushrooms only contain betulinic acid if they grow on birch trees. They are used in folk medicine as an anti-cancer treatment. Russian writer Aleksandr Solzhenitsyn came across chaga and its miraculous effects in the 1950s while researching his book Cancer Ward, published in 1968. In the novel, the main protagonist, Kostoglotov, is cured of stomach cancer mainly using chaga.[12] As well as cancer, chaga has been used to treat many other conditions including high or low blood pressure, diabetes, asthma, HIV, candidiasis, cardio-vascular disease, Crohn's disease and used as an anti-bacterial, anti-oxidant and anti-inflammatory agent. Chaga is often taken as a tea and an extract used as a food supplement.

Another birch fungus, *Piptoporus betulinus,* was carried by Ötzi, the Iceman, in his first-aid kit. This fungus contains toxic resins and agaric acid which are strong laxatives. Agaric acid was formerly used medicinally to stop perspiration. The fungus also contains anti-bacterial oils which are toxic to parasitic worms, like the whipworms Ötzi suffered from.[13] He used birch tar (and nettle fibres) to attach fletchings to the arrows he was carrying and also carried containers made of birch bark.[14]

References

1. http://www.aafo.com/goose/ [Accessed 13.12.2013]
2. J. Wright, (2012), http://www.theguardian.com/lifeandstyle/wordofmouth/2012/feb/01/how-to-make-birch-sap-wine [Accessed 13.12.2013]
3. http://www.imperial-leather.com/en-gb/our-story [Accessed 13.12.2013]
4. J. Lindahl, (2011); http://www.nordicwellbeing.com/Julies_Kitchen/2011/01/09/bark-bread-is-back/ [Accessed 12.12.2013]
5. C. Farrell; World Corporal Punishment Research, http://www.corpun.com/manx.htm [Accessed 13.12.2013]
6. http://www.bbc.co.uk/programmes/p01dgd9c/features/pollutionexperiment [Accessed 09.01.2013]

   and B.A. Maher, I.A.M. Ahmed, B. Davidson, V. Karloukovski, R. Clarke, (2013); *Environmental Science & Technology*, 47(23), 13737–44, doi: 10.1021/es404363m
7. Mrs M. Grieve, (1973 edition); *A Modern Herbal*, Merchant Book Company Ltd., ISBN 1904779018 pp 103–4
8. D. Hoffmann, (1996); *Complete Illustrated Guide to the Holistic Herbal*, Element Books, ISBN 0-00-713301-4, p 69
9. S. Fulda, (2008); *International Journal of Molecular Sciences*, 9(6), 1096–107, doi: 10.3390/ijms9061096
10. R.H. Cichewicz, S.A. Kouzi, (2004); *Medicinal Research Reviews*, 24(1), 90–114, doi: 10.1002/med.10053
11. J.-J. Tang, J.-G. Li, W. Qi, W.-W. Qiu, P.-S. Li, B.-L. Li, B.-L. Song, (2011); *Cell Metabolism*, 13(1), 44–56, doi: 10.1016/j.cmet.2010.12.004
12. R. Spinosa, (2006); *The Mycophile*; 47(1); http://namyco.org/images/pdf_files/MycoJanFeb06.pdf
13. L. Capasso, (1998), *The Lancet*, 352(9143), 1864, doi: 10.1016/S0140-6736(05)79939-6
14. A. Fleckinger, (2011); *Ötzi, the Iceman The Full Facts at a Glance*, 3rd updated edition, Folio, Vienna/Bolzano and the South Tyrol Museum of Archaeology, Bolzano, ISBN 978-3-85256-574-3, pp 86–7

December

18

19

20

21

22

23

24

December

25

26

27

28

29

30

31